SpringerBriefs in Molecular Science

Green Chemistry for Sustainability

Series Editor

Sanjay K. Sharma

For further volumes:
http://www.springer.com/series/10045

Peng Wu · Hao Xu · Le Xu
Yueming Liu · Mingyuan He

MWW-Type Titanosilicate

Synthesis, Structural Modification and Catalytic Applications to Green Oxidations

 Springer

Peng Wu
Hao Xu
Le Xu
Yueming Liu
Mingyuan He
Shanghai Key Laboratory of Green Chemistry
    and Chemical Processes
Department of Chemistry
East China Normal University
Shanghai
China

ISSN 2212-9898
ISBN 978-3-642-39114-9   ISBN 978-3-642-39115-6   (eBook)
DOI 10.1007/978-3-642-39115-6
Springer Heidelberg New York Dordrecht London

Library of Congress Control Number: 2013945287

Printed on acid-free paper

Springer is part of Springer Science+Business Media (www.springer.com)

Preface

The discovery of titanosilicates opens up new possibilities of developing hetero-geneous catalytic processes for selective oxidation reactions, which has made a breakthrough in the area of zeolite materials and catalysis. Although the first generation of titanosilicate TS-1 dates back to almost 30 years ago, the research activities are still being continued worldwide in design and synthesis of novel titanosilicates, insight into active sites as well as developing practically useful catalytic technologies. In this sense, a series of titanosilicates differing in crys-talline structure and pore dimension have been synthesized successfully. Particular efforts have been devoted to searching for the oxidation catalysts which have larger pore dimensions useful to process bulky molecules. Derived from lamellar precursors, so-called layered zeolites are constructing an important family in zeolite materials. Different from those with three-dimensional crystalline struc-tures already formed in hydrothermal synthesis, the layered zeolites possess structural diversity and their structures are mendable by post modification.

Focusing on recent research advances in a new generation of titanosilicate Ti-MWW that comes from a lamellar precursor, this monograph consists of five chapters. Chapter 1 introduces briefly the catalytic features and research progress of titanosilicate catalysts. Chapter 2 describes the methods for the preparation of Ti-MWW, including hydrothermal synthesis and post isomorphous substitution route either in the presence or absence of boric acid. Chapter 3 figures out the structural modifications of Ti-MWW, full or partial delamination, and interlayer pore expansion by, silylation or pillaring techniques. Chapter 4 deals with potential catalytic applications of thus developed catalysts to innovative selective oxidations including epoxidation of various alkenes and ammoximation of ketones to oxime. Chapter 5 gives the prospects for the development and application of Ti-MWW zeolite in future. The contents range from fundamental knowledge to practically usable techniques that have been established on this specific titanosilicate.

It is our great pleasure working in this research area with many excellent experts and students from both China and Japan. We would like to sincerely thank professors Takashi Tatsumi, Tatsuaki Yashima and Takayuki Komatsu

(Tokyo Institute of Technology, Japan), Osamu Terasaki (Stockholm University, Sweden) and Dr. Weibin Fan (Chinese Academy of Science, China) for their fruitful collaboration. Many students from East China Normal University, China and Yokohama National University, Japan have made great contributions to this research subject. Without their hard work and effort, it would have been impossible to put the independent topics together.

Shanghai, China

Peng Wu
Hao Xu
Le Xu
Yueming Liu
Mingyuan He

Contents

Chapter 1
Introduction

Zeolites are a class of crystalline aluminosilicates and silicalites with the silicon and aluminum cations tetrahedrally linked by the oxygen bridges in the framework, and then three-dimensional (3D) networks are constructed to form the channels, pores, cages, and cavities [1]. Zeolites thus possess well-defined crystalline structures as well as the textural properties of high specific surface area and high adsorption capacity. The pore windows and channels of zeolites are closely related to their well-defined crystalline structures, and usually their sizes are of molecular dimensions in the micropore region [2]. This kind of unique porosity endows zeolites with molecule sieving abilities for discriminating guest molecules and separating reactants/products. Even strong electric fields are possibly generated within zeolite pores, and as heterogeneous catalysts, zeolites then may exhibit strong quantum effects in combination with the molecular confinement of micropores [3, 4].

The chemical compositions of the zeolite frameworks are changeable and amendable, not only in the silicon to metal ratios but also in the types of coordinated metal ions. The Si/Al ratio could be varied in the range of one to infinite, whereas the transition metals and the elements other than Si and Al are also incorporated into the zeolite framework via isomorphous substitution, e.g., P [5], B [6], Ga [7], Fe [8], Ti [9], Sn [10], Ge [11], Zr [12], and V [13], etc., giving rise to so-called metallosilicates. This widens significantly the application range of zeolites as multifunctional catalysts.

Zeolites in aluminosilicate forms have long been used as solid–acid catalysts in petrochemical industry. For example, Faujasite Y (with a structure code of FAU, recognized by International Zeolite Association, IZA) is widely used as fluid catalytic cracking (FCC) catalyst, replacing conventional amorphous silica-alumina catalysts [14–16]. This opened the door for perhaps the biggest revolution in oil refining industry. Thereafter, the well-known pentasil aluminosilicate ZSM-5 with the MFI topology [17], developed in 1970s by Mobil (now known as ExxonMobil), is a versatile shape-selective catalyst in petrochemical processes for producing high valuable aromatics, e.g., para-xylene [18].

In 1983, Taramasso et al. [19] from Enichem group envisaged the first titanosilicate with the Ti cations isomorphously substituted in the MFI-type

P. Wu et al., *MWW-Type Titanosilicate*,
SpringerBriefs in Green Chemistry for Sustainability,
DOI: 10.1007/978-3-642-39115-6_1, © The Author(s) 2013

Fig. 1.1 Liquid-phase
oxidation reactions based on
TS-1/H$_2$O$_2$ system

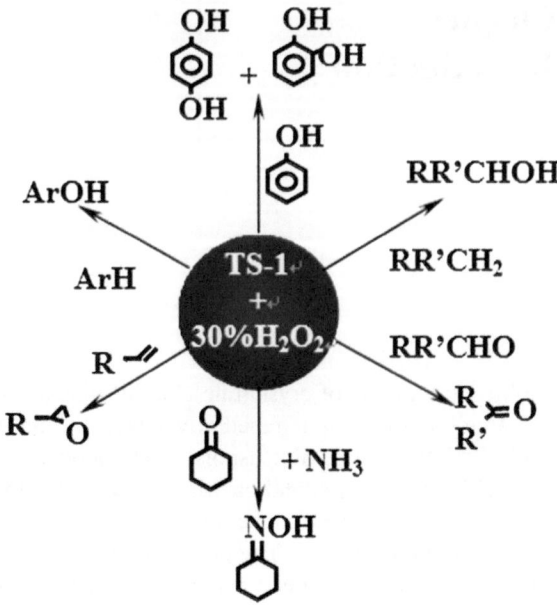

silicalite-1, which is well known as TS-1. Combining the hydrophobic feature in aqueous solution and the unique porosity of the MFI framework, those isolated tetrahedral Ti sites in TS-1 can activate hydrogen peroxide molecules under relatively mild conditions (generally <373 K and 1 atm), and as shown in Fig. 1.1, they are capable of catalyzing the selective oxidation reactions of a variety of substrates [20, 21]. These reactions give water as the sole byproduct. Two world-scale processes, cyclohexanone ammoximation [22] and propylene epoxidation [23], have now been commercialized based on the TS-1/H$_2$O$_2$ catalytic system. These processes are environmentally benign in terms of greenness and zero waste disposal. Expanding the catalytic applications of zeolites from solid–acid catalysis to redox field, the great success of TS-1 is considered as the third milestone in zeolite catalysis after Y and ZSM-5 zeolites.

The discovery of TS-1 has encouraged the researchers to develop other titanosilicates with different zeolite structures [24–49], especially those with larger porosities and high cost performance, because TS-1 encounters some shortcomings. TS-1 is less active to the bulky and cyclic molecules owing to diffusion hindrance imposed by its medium pores of 10-membered ring (MR). TS-1 always prefers a protic solvent like methanol, which causes the solvolysis of the epoxides in alkene epoxidation, lowering the selectivity to desirable products [50]. From the viewpoint of industrial applications, TS-1 still suffers a high cost of catalyst manufacturing because it requires the use of expensive tetrapropylammonium hydroxide (TPAOH) as structure-directing agent (SDA) and organic silica source of tetraethylorthosilicate (TEOS). To be active enough, the hydrothermally synthesized crystals of TS-1 needs to be of nanosized order. This then induces separation difficulties either in catalyst preparation procedure or in catalytic processes.

In this sense, taking full use of established zeolite preparation techniques such as hydrothermal synthesis (HT), post-synthesis (PS), dry-gel conversion (DGC) [51], fluoride method [52], crystallization-promoting agent or additive-assisted method [27], a series of titanosilicates have been developed. Table 1.1 lists the representative titanosilicates prepared by various techniques together with their crystalline and pore structures. In addition to microporous titanosilicates, many Ti-containing mesoporous materials with much larger nanopores have also been reported, such as Ti-MCM-41, Ti-SBA-15, and Ti-MCM-48 [53–56]. However, they are not effective for liquid-phase oxidation with aqueous H_2O_2 as oxidant because of an extremely high hydrophilicity related to abundant surface silanol groups.

Among the microporous titanosilicates developed so far, Ti-MWW with the MWW topology [33, 38, 57], has been proved to be unique in pore structure, preparation method, oxidation activity and selectivity. MWW zeolite possesses a unique pore structure of 12-MR side cups on the crystallites exterior as well as two independent 10-MR channel systems; one contains 12-MR supercages and the other is of sinusoidal tortuosity (Fig. 1.2) [58]. As the MWW zeolite derives from a so-called MWW lamellar precursor through the dehydroxylation between the layers upon calcination, it has structural diversity, e.g., conversion to a hybrid micro-mesoporous material by intercalating [59], fully or partially delamination into thin sheets with highly accessible external surface [60, 61], or structural transfer to interlayer-expanded zeolite structure by silylation with monomeric or dipodal silanes [62–64]. This would make MWW-based catalysts find much wider applications.

Although MWW aluminosilicate (generally known as MCM-22) is hydrothermally synthesized without difficulty, the synthesis of MWW titanosilicate (Ti-MWW) has been a challenge until we found that Ti could be effectively incorporated into the MWW framework when boric acid coexists with it in the synthesis media [33]. Making good use of structural characteristics of MWW zeolite, we have also established an original post-synthesis method for preparing boron-free Ti-MWW [38], and further converted it into a novel catalyst with more accessible active sites to bulky molecules. This book summarizes our recent works on preparing such novel titanosilicates and their catalytic properties in liquid-phase epoxidation of various alkenes as well.

Table 1.1 List of representative titanosilicates: crystalline structure, pore system and synthesis methods

Titanosilicate	Structure code	Pore channels (MR)	Synthesis method	Framework	References
TS-1	MFI	10-10	HTS, PS		[19]
TS-2	MEL	10-10-10	HTS		[24]
Ti-Beta	*BEA	12-12-12	HTS, F⁻		[25–29]
Ti-MOR	MOR	12-8	PS		[30–32]
Ti-MWW	MWW	10-10 12 supercage	HTS, PS, DGC, F⁻		[33–40]
Ti-ZSM-48	*MRE	10	HTS		[41]
Ti-FER	FER	10-8	HTS		[42]

(continued)

Table 1.1 (continued)

Titanosilicate	Structure code	Pore channels (MR)	Synthesis method	Framework	References
TAPSO-5	AFI	12	HTS		[43]
Ti-ZSM-12	MTW	12	HTS		[44]
Ti-MCM-68	MSE	12-10-10	PS		[45]
Ti-ITQ-7	ISV	12-12-12	HTS		[46, 47]
Ti-UTD-1	DON	14	HTS		[48]
Ti-CDS-1	CDO	10-8	HTS		[49]

HTS hydrothermal synthesis, *DGC* dry gel conversion, *PS* postsynthesis, F^- fluoride media method

Fig. 1.2 Topology of MWW
zeolites (*h0l* plane)

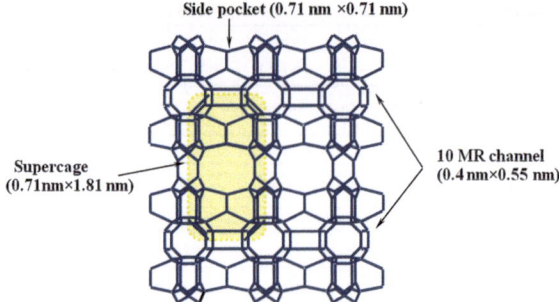

Side pocket (0.71 nm ×0.71 nm)

Supercage
(0.71nm×1.81 nm)

10 MR channel
(0.4 nm×0.55 nm)

References

1. Moscou L (1991) Introduction to zeolite science and practice. In: Van Bekkum H, Flanigen EM, Jacobs PA, Jansen JC (eds) Stud. Surf. Sci. Catal., vol 1. Elsevier, Amsterdam, p 58
2. Davis ME, Lobo RF (1992) Zeolite and molecular sieve synthesis. Chem Mater 4:756–768
3. Sastre G, Corma A (2009) The confinement effect in zeolites. J Mol Catal A- Chem 305:3–7
4. Zicovich CM, Corma A (1994) Electronic confinement of molecules in microscopic pores. A new concept which contributes to explanation of the catalytic activity of zeolites. J Phys Chem 98:10863–10870
5. Lok BM, Messina CA, Patton RL et al (1984) Silicoaluminophosphate molecular sieves: another new class of microporous crystalline inorganic solids. J Am Chem Soc 106:6092–6093
6. Millini R, Perego G, Parker WO et al (1995) Layered structure of ERB-1 microporous borosilicate precursor and its intercalation properties towards polar molecules. Micropor Mater 4:221–230
7. Thomas JM, Liu XS (1986) Gallozeolite catalysts: preparation, characterization and performance. J Phys Chem 90:4843–4847
8. Lewis DW, Catlow CRA, Sankar G et al (1995) Structure of iron-substituted ZSM-5. J Phys Chem 99:2377–2383
9. Wu P, Liu Y, He M et al (2004) A novel titanosilicate with MWW structure: catalytic properties in selective epoxidation of diallyl ether with hydrogen peroxide. J Catal 228:183–191
10. Corma A, Nemeth LT, Ren M et al (2001) Sn-zeolite beta as a heterogeneous chemoselective catalyst for Baeyer-Villiger oxidations. Nature 412:423–425
11. Castaneda R, Corma A, Fornes V et al (2003) Synthesis of a new zeolite structure ITQ-24, with intersecting 10- and 12- member ring pores. J Am Chem Soc 125:7820–7821
12. Nie Y, Jaenicke S, Chuah G-K (2009) Zr-zeolite Beta: a new heterogeneous catalyst system for the highly selective cascade transformation of citral to (±)-menthol. Chem Eur J 15:1991–1999
13. Cavani F, Trifiro F, Jirrů P et al (1988) Vanadium-zeolite catalysts for the ammoxidation of xylenes. Zeolites 8:12–18
14. Bevilacqua M, Montanari T, Finocchio E et al (2006) Are the active sites of protonic zeolites generated by the cavities? Catal Today 116:132–142
15. Zečević J, Gommes CJ, Friedrich H et al (2012) Mesoporosity of zeolite Y: quantitative three-dimensional study by image analysis of electron tomograms. Angew Chem Int Ed 51:4213–4217
16. Janssen AH, Koster AJ, de Jong KP (2001) Three-dimensional transmission electron microscopic observations of mesopores in dealuminated zeolite Y. Angew Chem Int Ed 40:1102–1104

17. Kokotailo GT, Lawton SL, Olson DH et al (1978) Structure of synthetic zeolite ZSM-5. Nature 272:437–438
18. Beschmann K, Riekert L (1993) Isomerization of xylene and methylation of toulene on zeolite H-ZSM-5. Compound kinetics and selectivity. J Catal 141:548–565
19. Taramasso M, Perego G, Notari B (1983) US 4410501
20. Bellussi G, Rigutto MS (2001) Metal ions associated to molecular sieve frameworks as catalytic sites for selective oxidation reactions. Stud Surf Sci Catal 137:911–955
21. Ratnasamy P, Srinivas D, Knözinger H (2004) Acitve sites and reactive intermediates in titanium silicate molecular sieves. Adv Catal 48:1–169
22. Roffia P, Leofanti G, Cesana A et al (1990) Cyclohexanone ammoximation: a break through in the 6-caprolactam production process. Stud Surf Sci Catal 55:43–52
23. Wells DH Jr, Delgass WN, Thomson KT (2004) Evidence of defect-promoted reactivity for epoxidation of propylene in titanosilicate (TS-1) catalysts: a DFT study. J Am Chem Soc 126:2956–2962
24. Reddy JS, Kumar R, Ratnasamy P (1990) Titanium silicate-2: synthesis, characterization and catalytic properties. Appl Catal 58:L1–L2
25. Camblor MA, Corma A, Martínez A et al (1992) Synthesis of a titaniumsilicoaluminate isomorphous to zeolite beta and its application as a catalyst for the selective oxidation of large organic molecules. J Chem Soc, Chem Commun 8:589–590
26. Camblor MA, Constantini M, Corma A et al (1996) Synthesis and catalytic activity of aluminium-free zeolite Ti-β oxidation catalysts. Chem Commun 11:1339–1340
27. Blasco T, Camblor MA, Corma A et al (1996) Unseeded synthesis of Al-free Ti-β zeolite in fluoride medium: a hydrophobic selective oxidation catalyst. Chem Commun 20:2367–2368
28. Jappar N, Xia Q, Tatsumi T (1998) Oxidation activity of Ti-Beta synthesiszed by a dry-gel conversion method. J Catal 180:132–141
29. Tatsumi T, Jappar N (1998) Properties of Ti-Beta zeolites synthesized by dry-gel conversion and hydrothermal methods. J Phys Chem B 102:7126–7131
30. Wu P, Komatsu T, Yashima T (1996) Characterization of titanium species incorporated into dealuminated mordenites by means of IR spectroscopy and ^{18}O-Exchange technique. J Phys Chem 100:10316–10322
31. Wu P, Komatsu T, Yashima T (1997) Ammoximation of Ketones over titanium mordenite. J Catal 168:400–411
32. Wu P, Komatsu T, Yashima T (1998) Hydroxylation of aromatics with hydrogen peroxide over titanosilicates with MOR and MFI structures: effect of Ti peroxo species on the diffusion and hydroxylation activity. J Phys Chem B 102:9297–9303
33. Wu P, Tatsumi T, Komatsu T, Yashima T (2001) A novel titanosilicate with MWW structure. I. Hydrothermal synthesis, elimination of extraframework titanium, and characterizations. J Phys Chem B 105:2897–2905
34. Wu P, Tatsumi T, Komatsu T, Yashima T (2001) A novel titanosilicate with MWW structure: II. Catalytic properties in the selective oxidation of alkenes. J Catal 202:245–255
35. Wu P, Tatsumi T (2001) Extremely high *trans* selectivity of Ti-MWW in epoxidation of alkenes with hydrogen peroxide. Chem Commun 10:897–898
36. Wu P, Tatsumi T (2002) Unique *trans*-selectivity of Ti-MWW in epoxidation of *cis/trans*-alkenes with hydrogen peroxide. J Phys Chem B 106:748–753
37. Wu P, Tatsumi T (2003) A novel titanosilicate with MWW structure. III. Highly efficient and selective production of glycidol through epoxidation of allyl alcohol with H$_2$O$_2$. J Catal 214:317–326
38. Wu P, Tatsumi T (2002) Preparation of B-free Ti-MWW through reversible structural conversion. Chem Commun 10:1026–1027
39. Fan WB, Wu P, Namba S et al (2004) A titanosilicate that is structurally analogous to an MWW-Type lamellar precursor. Angew Chem Int Ed 43:236–240
40. Fan WB, Wu P, Namba S et al (2006) Synthesis and catalytic properties of a new titanosilicate molecular sieve with the structure analogous to MWW-type lamellar precursor. J Catal 243:183–191

41. Serrano DP, Li HX, Davis ME (1992) Synthesis of titanium-containing ZSM-48. J Chem Soc, Chem Commun 10:745–747
42. Ahedi RK, Kotasthane AN (1998) Synthesis of FER titanosilicates from a non-aqueous alkali-free seeded system. J Mater Chem 8:1685–1686
43. Tuel A (1995) Synthesis, characterization, and catalytic properties of titanium silicoaluminophosphate TAPSO-5. Zeolites 15:228–235
44. Tuel A (1995) Synthesis, characterization, and catalytic properties of the new TiZSM-12 zeolite. Zeolites 15:236–242
45. Kubota Y, Koyama Y, Yamada T et al (2008) Synthesis and catalytic performance of Ti-MCM-68 for effective oxidation reactions. Chem Commun 46:6224–6226
46. Díñaz-Cabañas MJ, Villaescusa LA, Camblor MA (2000) Synthesis and catalytic activity of Ti-ITQ-7: a new oxidation catalyst with a three-dimensional system of large pore channels. Chem Commun 9:761–762
47. Corma A, Díñaz-Cabañas MJ, Domine E et al (2000) Ultra fast and efficient synthesis of Ti-ITQ-7 and positive catalytic implications. Chem Commun 18:1725–1726
48. Balkus KJ Jr, Gabrielov AG, Zones SI (1995) The synthesis of UTD-1, Ti-UTD-1 and Ti-UTD-8 using CP*$_2$CoOH as a structure directing agent. Stud Surf Sci Catal 97:519–525
49. Wu L, Liu Y, Wu P et al (2008) Hydrothermal synthesis and characterization of CDS-1 type titanosilicate. Acta Chim Sinica 66:141–144
50. Fan W, Wu P, Tatsumi T (2008) Unique solvent effect of microporous crystalline titanosilicates in the oxidation of 1-hexene and cyclohexene. J Catal 256:62–73
51. Xu W, Dong J, Li J et al (1990) A novel method for the preparation of zeolite ZSM-5. J Chem Soc, Chem Commun 10:755–756
52. Vidal-Moya JA, Blasco T, Rey F et al (2003) Distribution of fluorine and germanium in a new zeolite structure ITQ-13 studied by ^{19}F nuclear magnetic resonance. Chem Mater 105:3961–3963
53. Blasco T, Corma A, Navarro MT et al (1995) Synthesis, characterization, and catalytic activity of Ti-MCM-41 structures. J Catal 156:65–74
54. Newalkar BL, Olanrewaju J, Komarneri S (2001) Direct synthesis of titanium-substituted mesoporous SBA-15 molecular sieve under microwave–hydrothermal conditions. Chem Mater 13:552–557
55. Wu P, Tatsumi T, Komatsu T et al (2002) Postsynthesis, characterization, and catalytic properties in alkene epoxidation of hydrothermally stable mesoporous Ti-SBA-15. Chem Mater 14:1657–1664
56. Koyano KA, Tatsumi T (1996) Synthesis of titanium-containing mesoporous molecular sieves with a cubic structure. Chem Commun 2:145–146
57. Liu N, Liu YM, Xie W et al (2007) Hydrothermal synthesis of boron-free Ti-MWW with dual structure-directing agents. Stud Surf Sci Catal 170:464–469
58. Leonowicz ME, Lawton JA, Lawton SL et al (1994) MCM-22: a molecular sieve with two independent multidimensional channel systems. Science 264:1910–1913
59. Kim SY, Ban HJ, Ahn WS (2007) Ti-MCM-36, a new mesopore epoxidation catalyst. Catal Lett 113:160–164
60. Wu P, Nuntasri D, Ruan J et al (2004) Delamination of Ti-MWW and high efficiency in epoxidation of alkenes with various molecular sizes. J Phys Chem B 108:19126–19131
61. Wang L, Wang Y, Liu Y et al (2008) Post-transformation of MWW-type lamellar precursors into MCM-56 analogues. Micropro Mesopro Mater 113:435–444
62. Wu P, Ruan J, Wang L et al (2008) Methodology for synthesizing crystalline metallosilicates with expanded pore windows through molecular alkoxysilylation of zeolitic lamellar precursors. J Am Chem Soc 130:8178–8187
63. Wang L, Wang Y, Liu Y et al (2009) Alkoxysilylation of Ti-MWW lamellar precursors into interlayer pore-expanded titanosilicates. J Mater Chem 19:8594–8602
64. Corma A, Diaz U, Garcia T et al (2010) Multifunctional hybrid organic-inorganic catalytic materials with a hierarchical system of well-defined micro- and mesopores. J Am Chem Soc 132:15011–15021

Chapter 2
Synthesis of Ti-MWW Zeolite

Abstract Titanosilicate with MWW topology can be prepared by hydrothermal method, dry-gel method, and post-synthesis method. At first, the addition of B atoms in the synthesis gels was identified as the key factor for the successful synthesis of Ti-MWW under hydrothermal condition. Then, several alternative methods, for example, hydrothermal method with dual SDAs and post-synthesis method, were developed to provide B-less or B-free condition, which avoided a waste of B atoms as well as the framework acidity introduced by B atoms. Ti-MWW zeolites prepared by different methods varied in particle size and activity in liquid oxidation reactions, for example, the epoxidation of 1-hexene.

Keywords Ti-MWW · Synthesis · B content · Post-synthesis method · Epoxidation

2.1 Introduction

The first titanosilicate TS-1 has achieved great success in the area of redox catalysis [1, 2]. It is not only highly efficient but also zero waste disposal, several zeolite structures with larger pores have been chosen as candidates for synthesizing titanosilicates such as MOR [3], BEA [4], MTW [5]. The purpose is to overcome the deficiencies that the medium-pore TS-1 catalyst meets with less activity in the oxidation involving the substrates with large molecular dimensions. Post-synthesized Ti-MOR with one-dimension 12-MR channels has been proved to be highly active in the ammoximation of ketones [6] and the hydroxylation of dutrex [7]. However, Ti-MOR is almost inactive in the epoxidation of alkenes. Although Ti-MOR zeolite possesses larger-pore channels than TS-1, their channels are not inter-connective [8]. This makes it lack the potential activity possessed by the zeolites with three-dimensionally (3D) connected channels. Ti-Beta with a 3D

P. Wu et al., *MWW-Type Titanosilicate*,
SpringerBriefs in Green Chemistry for Sustainability,
DOI: 10.1007/978-3-642-39115-6_2, © The Author(s) 2013

12-MR pore system shows advantages in the epoxidation of cyclic alkenes with H_2O_2, but it is intrinsically less active than TS-1 in the reactions involving the small-size substrates, which is due to the large amount of defects in the Beta structure [9]. Moreover, the leaching of Ti species during the liquid phase reactions is very obvious for Ti-Beta zeolite. Apart from microporous zeolites, Ti-containing mesoporous zeolites such as Ti-MCM-41[10] have also been prepared and they showed high activity only when *tert*-butyl hydroperoxide (TBHP) was adopted as the oxidant. The low mechanical and hydrothermal stability, easy leaching of Ti species in liquid phase reactions, and lower intrinsic activity make Ti-MCM-41 less active than TS-1 when the substrates do not suffer diffusion hindrance problems.

MWW-type zeolite, with two independent set of 10-MR channels, was first constructed from a lamellar precursor by calcination, during which dehydration and condensation take place to form 3D MWW structure [11]. Those diffraction peaks related to the layer stacking direction, *c*-direction, would disappear upon calcination while other diffraction peaks due to the MWW sheets are almost intact. Aluminum containing MWW-type zeolite, well known as MCM-22, has shown great success in the process of benzene alkylation as a solid acid catalyst since it was found, mainly owing to its unique structure [12]. In addition to two independent sets of 10-MR channels, the MWW structure possesses also 12-MR supercages $(0.7 \times 0.7 \times 1.8$ nm), which turn to be pockets or cup moieties $(0.7 \times 0.7$ nm) on the crystal outer surface. The intracrystalline supercages and exterior pockets covering the hexagonal flake-like crystals are considered to serve as the open reaction spaces for the disproportionation of toluene and alkylation of benzene. Considering that the MWW-type structure is stable and unique, it is supposed to show unusual activity as a redox catalyst when the transition metal cations are incorporated into the framework. Unfortunately, the road for preparing Ti containing MWW-type (Ti-MWW) zeolite was not as smooth as that of trivalent cations containing MWW zeolites such as Al-MWW [13], B-MWW [14] and Fe-MWW [15].

Ti-containing zeolites are prepared mainly by direct hydrothermal method [16], post-synthesis method [17], and dry-gel conversion (DGC) method [18]. Different from the aluminosilicates, the synthetic system for preparing titanosilicates with high activity should contain as small amount of other metal ions as possible, which means a high siliceous synthetic gel is preferred for preparing Ti-containing zeolites. However, some zeolite structures such as MOR zeolite cannot be constructed in siliceous gels if without the help of Al^{3+} and Na^+ cations [19]. Post-synthesis method was then employed as another effective way to prepare those titanosilicates which cannot be directly synthesized by the hydrothermal method. In this sense, Ti-MOR was successfully post-synthesized using a so-called "atom-planting" route, which involved the gas–solid reaction between highly dealuminated MOR zeolite and $TiCl_4$ vapor at elevated temperatures [6]. Grafting technique is also a useful post-synthesis method especially in preparing Ti-containing Mesoporous materials such as Ti-MCM-41[20] and Ti-SBA-15 [21]. The DGC method involves the dehydration of synthetic gels and the crystallization assisted by steam that is placed separately from the synthetic gel powder. The DGC method has achieved a great success in preparing Ti-Beta with less Al^{3+} content and smaller crystal size than conventional hydrothermal synthesis [22].

Hydrothermal synthesis is the most widely used method to prepare zeolites including both pure silicate zeolites and metal-doped zeolites. However, it is difficult to obtain Ti-MWW from the synthetic gels only containing silicon and titanium via hydrothermal method. Since the direct synthesis of Ti-MWW was not so easy, the post-synthesis method became an alternative. The first success of the titanosilicate with the MWW topology is Ti-ITQ-2 [23], which was prepared by grafting titanocene onto the surface of delaminated ITQ-2 silicalite. ITQ-2, containing abundant silanols exposed on the surface, is a delaminated form of the MWW structure. Nevertheless, Ti-ITQ-2 zeolite shows high activity only when using TBHP as an oxidant, showing very similar catalytic behaviors to mesoporous titanosilicates such as Ti-MCM-41. It is the hydrophilic character that leads to a low activity of Ti-ITQ-2 in H_2O_2 system and Ti leaching during liquid-phase reactions. Atom-planting technique is another post-synthesis method adopted to prepare Ti-MWW zeolite, in which gas phase $TiCl_4$ was used as Ti source to react with dealuminated Al-MCM-22 zeolite [24]. However, severe leaching of Ti species occurs in the Ti-MWW zeolite obtained by atom-planting method. Although there are so many difficulties in the synthesis of Ti-MWW zeolite, the above efforts are not in vain but pave the way for the following study.

The breakthrough in the synthesis of Ti-MWW zeolite with both high crystallinity and activity was the introduction of boric acid as structure-supporting agent, which was based on the synthesis of B-containing MWW zeolite ERB-1 [25]. ERB-1 zeolite can be synthesized both in the presence and the absence of alkali cations, which means B^{3+} cations can serve as structure-supporting agent. Borosilicate with the MWW topology can be obtained from the gels with Si/B = 1.5. However, Ti-MWW zeolite cannot be synthesized in the gels with Si/B = 1.5 unless a large amount of boron cations were introduced with Si/B = 0.75 [26]. Although the acidity of B^{3+} ions is much weaker than that of Al^{3+} ions, the introduction of B atom into the MWW framework would inevitably increase the electronegativity and further affect the performance of the Ti species in catalysis. Hence, several other methods (Table 2.1) such as F^- ions assisted method, DGC method [27] and post-synthesis method [28] were also employed to prepare Ti-MWW zeolite with the aim of decreasing the B content. Ti-MWW zeolites prepared with different methods varied in particle sizes and activity. These methods can be divided into three kinds based on the B content in synthetic gels: B-containing method, B-less method, and B-free method. All the synthesis methods are explained in detail in following text.

Table 2.1 A summary of synthesis methods for Ti-MWW[a]

No.	Si source	SDA[b]	Method[c]	Gel composition		Crystal size (μm)[d]		Ref.
				Si/B	Si/Ti			
1	TEOS	PI, HMI	HTS	0.75	10 ~ ∞	PI	(0.2–0.5)* (0.05–0.1)	[26]
						HMI	1*0.1	
2	Fumed silica	OCTMAOH HEPMAOH HEXMAOH	HTS	5–10	30 ~ ∞		0.15*0.02	[41]
3	Fumed silica	TMAadOH and HMI	HTS	∞	20 ~ ∞		0.5*0.1	[57]
4	Fumed silica Colloidal silica	PI, HMI	DGC	1–92	30 ~ ∞		4–8	[27]
5	TEOS Fumed silica	PI, HMI	F⁻	1–15	15 ~ ∞		2*0.2	–
6	Siliceous MWW	PI, HMI	PS	∞	20 ~ ∞	PI	(0.2–0.5)* (0.05–0.1)	[28]
						HMI	1*0.1	

[a] The Ti source used in all the syntheses is tetrabutylorthotitanate (TBOT)

[b] *PI* piperidine, *HMI* hexamethyleneimine, *OCTMAOH* octyltrimethylammonium hydroxide, *HEPMAOH* heptyltrimethylammonium hydroxide, *HEXMAOH* hextyltrimethylammonium hydroxide, *TMAadOH* N,N,N-trimethyl-1-adamantanammonium hydroxide

[c] *HTS* hydrothermal synthesis, *DGC* dry gel conversion; *F⁻* crystalized with the assistant of F⁻ ions, *PS* postsynthesis

[d] Crystal size is presented as the length multiplied by the thickness

2.2 Boron-Containing Ti-MWW Synthesis

With the assistance of boric acid, Ti-MWW zeolite (denoted as Ti-MWW-HTS) was hydrothermally synthesized for the first time in a series of Ti-content with the Si/Ti ratios ranging from 10 to ∞. It can be seen from Table 2.2 that only a small portion of the B^{3+} cations in the synthetic gels was incorporated into the framework of Ti-MWW zeolite, which means that most of the B^{3+} cations were useless in constructing the MWW structure [26]. The case was exactly the contrary as for Ti^{4+} cations. Ti^{4+} cations added in the synthetic gels were all incorporated into the Ti-MWW-HTS zeolite, although the crystallinity was slightly decreased with the increase of Ti content. The insertion of Ti^{4+} into the framework has been proved by both UV–Vis and IR measurements.

Two adsorption bands were observed in the UV–visible spectra (Fig. 2.1), locating around 220 and 260 nm, respectively, which indicate there are two kinds of Ti species in the as-synthesized Ti-MWW zeolite. The band around 220 nm is due to the tetrahedral Ti species [29] in the framework of Ti-MWW-HTS zeolite while the band around 260 nm is attributed to the octahedral Ti species [30] on the external surface of the zeolite. Irrespective of a high Ti content like Si/Ti ratio of 20 in the gel, anatase phase was never detected, indicating the Ti species were highly dispersed. The presence of 260 nm adsorption-related octahedral species is

Table 2.2 Hydrothermal synthesis of Ti-MWW from different gel compositions

No.	Gel composition[a]		Product composition and surface area					
	Si/B	Si/Ti	Ti-MWW-PI			Ti-MWW-HMI		
			Si/B	Si/Ti	SSA[b] (m^2 g^{-1})	Si/B	Si/Ti	SSA[b] (m^2 g^{-1})
1	0.75	∞	11.8	∞	616	13.6	∞	601
2	0.75	100	12.6	120	625	16.3	138	621
3	0.75	70	12.2	63	612	14.2	79	628
4	0.75	50	11.4	51	621	12.4	53	–
5	0.75	30	11.0	31	623	11.6	31	613
6	0.75	20	12.7	21	540	11.4	22	–[c]
7	0.75	10	13.6	10	537	11.5	9.6	541

Reprinted from Ref. [26], Copyright 2001, with permission from American Chemical Society
[a] Other gel compositions: PI or HM/SiO$_2$ = 1.4; H$_2$O/SiO$_2$ = 19
[b] SSA specific surface area (Langmuir)
[c] Not determined

simply because as-synthesized Ti-MWW-HTS has a lamellar precursor structure. Those Ti species introduced on the layer surface tend to occupy the octahedral coordination. This is an unusual aspect greatly different from those titanosilicates which have the 3D crystalline structures already in as-synthesized form.

Both piperidine (PI) and hexamethyleneimine (HMI) serve as the structure-directing agents (SDA) in preparing Ti-MWW-HTS zeolite, denoted as Ti-MWW-HTS-PI and Ti-MWW-HTS-HMI, respectively. They possess almost the same surface area (Table 2.2), but the particle size of Ti-MWW-HTS-HMI is larger than Ti-MWW-HTS-PI, which induces the difference of Ti^{4+} distribution on the crystal of Ti-MWW zeolite. As a result, Ti-MWW-HTS-PI with smaller particle size and larger external surface area possesses more Ti species on the external surface than Ti-MWW-HTS-HMI, which induces a main adsorption band around 260 nm in UV–Vis spectra. As for Ti-MWW-HTS-HMI, the main adsorption band changes from 220 to 260 nm with the increase of Ti content, which means Ti^{4+} ions take place the intracrystalline position in priority. When the as-synthesized Ti-MWW-HTS zeolites were suffered with calcination treatment, a new band around 330 nm appeared on both the UV–visible spectra of Ti-MWW-HTS-PI and Ti-MWW-HTS-HMI, indicating an anatase phase [31] formed probably due to the partial condensation and aggregation of neighboring surface Ti species upon calcination. These extraframework Ti species should be removed because it would cause the unproductive decomposition of hydrogen peroxide in actual reactions, which would certainly lead to a poor performance of Ti-MWW zeolite as an oxidation catalyst. An acid treatment was employed to remove the inactive Ti species in the calcined form of Ti-MWW-HTS zeolite. As indicated by the UV–visible spectra, the acid treatment decreased slightly the bands around 220 and 260 nm, while it hardly affected the band around 330 nm. This revealed that the anatase phase can withstand the acid treatment and hardly be removed from the Ti-MWW-HTS zeolite once the anatase phase formed during calcination.

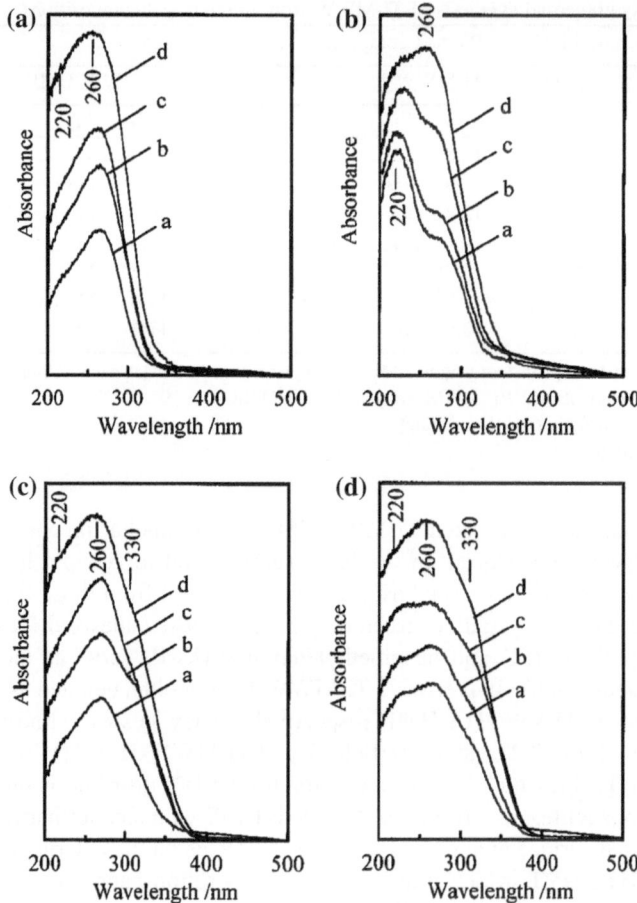

Fig. 2.1 UV–visible spectra of (**a**) as synthesized and (**c**) calcined Ti-MWW-PI, and (**b**) as-synthesized and (**d**) calcined Ti-MWW-HMI with the Si/Ti ratio of (*a*) 100, (*b*) 50, (*c*) 30, (*d*) 10. Reprinted from Ref. [26], Copyright 2001, with permission from American Chemical Society

Then the acid extraction was tried on the as-synthesized Ti-MWW-HTS zeolite with the purpose of removing the anatase phase from the precursor in advance of the calcination treatment. The XRD pattern of the acid treated Ti-MWW-HTS zeolite was almost the same as that of calcined Ti-MWW-HTS zeolite, with 001 and 002 diffractions disappeared, which was probably due to the removal of support such as Ti and organic molecules between the layers. UV–visible spectra verified that the octahedral Ti species on the external surface of Ti-MWW crystals were removed selectively upon the acid treatment. Further calcination made the diffraction peaks more intense and resulted in an anatase-free Ti-MWW-HTS zeolite as verified by the UV–visible spectra again. It should be pointed out that the octahedral Ti species cannot be fully removed by acid treatment in the case of

Ti-MWW-HTS lamellar precursors with Si/Ti ratio lower than 20. Nevertheless, the tetrahedral Ti species were almost intact in the framework during acid treatment procedure. In such an acid and subsequent calcination procedure, the inactive Ti species were almost removed and the B content was also decreased a lot. If a further acid treatment was adopted, a nearly B-free Ti-MWW zeolite was obtained while the Ti content was almost unchanged during the second round of acid treatment. As shown in Fig. 2.2, Ti-MWW nearly free of both B and anatase can be synthesized with PI or HMI as SDA under alkali-free conditions, followed by a

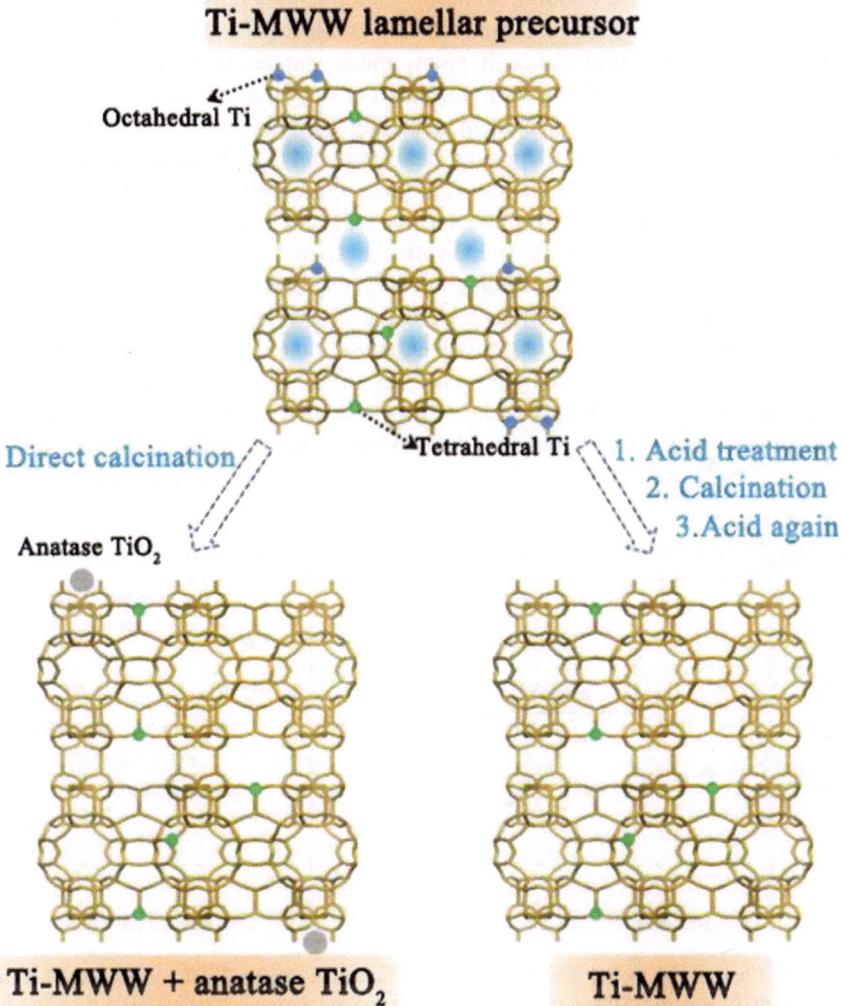

Fig. 2.2 MWW-type titanosilicate containing only tetrahedral Ti species in the framework was prepared by a cyclic acid treatment

cyclic post-treatment, that is, an acid treatment, subsequent calcination and a further acid treatment.

A detailed study on the performance of Ti-MWW-HTS zeolite in the epoxidation of alkenes revealed that Ti-MWW-HTS showed a superior activity independent of the oxidant in comparison with microporous TS-1 and Ti-MOR and mesoporous Ti-MCM-41 [32]. As compared with Ti-Beta zeolite, Ti-MWW showed a better performance in the epoxidation of cyclohexene using TBHP as the oxidant. However, Ti-Beta zeolite showed a higher activity in the case of H_2O_2 as the oxidant. The stabilities of both structure and active species are most concerned factors in judging a catalyst. The Ti content was almost unchanged in the liquid-phase cyclohexene epoxidation, while the B content decreased nearly half of the fresh catalyst after the first use and further decreased gradually with the reaction–regeneration cycles (Fig. 2.3a). B atoms much smaller than silicon atoms are easily cleaved from the framework, forming defects simultaneously. Two kinds of regeneration treatment were applied including washing with acetone and calcination (Fig. 2.3b). The activity was recovered only 75 % upon washing with acetone while the turnover number (TON) was totally restored by calcination. The severe leaching of B atoms would leave vacancy and then cause the Ti migration upon the rearrangement of the framework occurring in the vicinity of B cations. The regeneration treatment of calcination was supposed to mend those defects and then to stabilize the framework as well as the Ti species.

As mentioned above, B atoms involved in the framework was expected to have weaker influence on the catalytic behaviors of the Ti sites than the Al atoms. A detailed study of B effect was investigated in the epoxidation of cyclohexene, which revealed that the B content exhibited almost no influence on the conversion

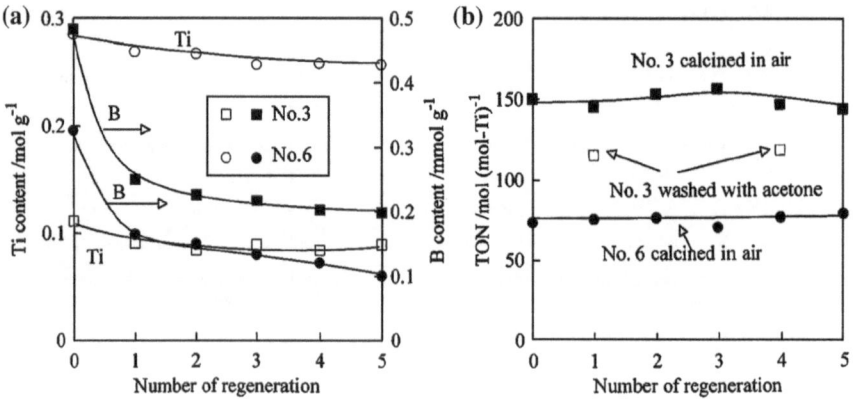

Fig. 2.3 Changes of the Ti and B contents (**a**) and that of the TON for the cyclohexene conversion (**b**) with the cycle of reaction–regeneration. Cyclohexene oxidation: 0.05 g Ti-MWW, 10 mmol of substrate, 5 mmol of H_2O_2, 5 mL of acetonitrile; temp., 333 K; time, 2 h. *No. 3* indicates Ti-MWW-PI(100) treated by 2 M HNO_3, while *No. 6* indicates Ti-MWW-PI(70) treated by 6 M HNO_3. Reprinted from Ref. [32], Copyright 2001, with permission from Elsevier

of cyclohexene but increased the selectivity to ring-opening products. The existence of B atoms in the framework of Ti-MWW zeolite would not poison the Ti active sites but arouse the ring-opening and solvolysis reactions of the target epoxide product due to the weak acidity related to the Si(OH)B groups.

The conversion of 1-hexene increased with the increase of Ti content, while the TON value showed a volcanic shape change (Fig. 2.4). On the contrary, the TON value decreased with the increase of Ti content when the epoxidation of cyclohexene was carried out over Ti-MWW zeolite. The two different variations were believed to be related to the molecular size and the distribution of Ti species in the Ti-MWW crystals. Ti species can be located in three kinds of sites in the Ti-MWW zeolite, that is, the 10-MR channels, the intracrystalline supercages, and the external surface 12-MR pockets. 1-Hexene molecules were small enough to diffuse into the 10-MR channels and access all the Ti species, whereas the cyclohexene molecules can only reach the supercages and the outer surface pockets. With the increase of Ti content, the Ti atoms would occupy the intracrystalline positions in priority then the outer surface, which has been proved by the UV–visible spectra [26]. When the Si/Ti ratio was higher than 40, the Ti species were almost located in the inner space of the Ti-MWW crystalline, where the larger molecules are less easily to access to. As a result, the TON values were decreased as for cyclohexene and increased as for 1-hexene with the increase of Ti content. However, when the octahedral Ti species appeared with an extreme excess amount, it is impossible to remove them completely by acid treatment, certainly causing a decrease of the TON values in both cyclohexene and 1-hexene epoxidation when the Si/Ti ratio was lower than 40.

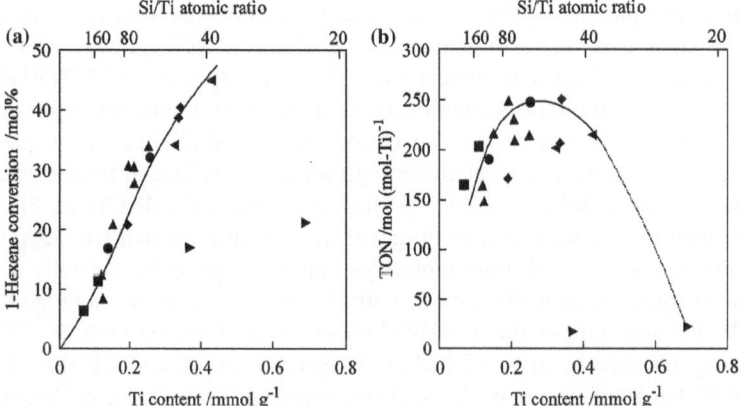

Fig. 2.4 Dependence of the conversion (**a**) and the TON (**b**) on the Ti content for the oxidation of 1-hexene with H_2O_2. Ti–MWW–PI catalysts used were prepared from the lamellar precursors with a Si/Ti ratio of 100 (*filled square*), 70 (*filled circle*), 50 (*filled up-pointing triangle*), 30 (*filled diamond*), 20 (*filled left-pointing triangle*), and 10 (*filled right-pointing triangle*). Reprinted from Ref. [32], Copyright 2001, with permission from Elsevier

2.3 Boron-Less Ti-MWW Synthesis

2.3.1 Synthesis of Ti-MWW Using Linear-Type Quaternary Alkylammonium Hydroxides

Organic SDAs play important roles in the nucleation and crystallization process of the zeolite synthesis, including compensating the structural electronegativity, filling the pore channels to stabilize the whole zeolite structure and interacting with the inorganic species to direct the formation of zeolite structures [33, 34]. All the MWW analogues are commonly synthesized with cyclic or polycyclic amine [26]. HMI was used to synthesize PHS-3 [35] and MCM-22 [26] zeolites under alkali conditions, while a special organic cations N,N,N-trimethyl-1-adamantylammonium (TMAda$^+$) was employed to construct SSZ-25 [36, 37] with the help of K$^+$ cations. These organic molecules with large molecular sizes seem to prefer to occupy the interlayer space rather than stabilize the zeolite structure. On the other hand, linear-type organic ammonium molecules have also been used to prepare zeolite with MWW topology, such as N,N,N,N',N',N'-hexamethyl-1,5-pentanediammonium and 1,4-bis(N-methylpyrrodinium)butane, although their structure-directing ability is relatively weak [38, 39]. The linear-type organic additives with relatively smaller sizes are more flexible than cyclic molecules and may fill both the interlayer space and the intralayer 10-MR channels in MWW structure. Liu et al. found that a linear quaternary ammonium, octyltrimethylammonium hydroxide (OCTMAOH), had the ability to direct borosilicate and titanosilicate with MWW topology under alkali-free conditions [40]. MWW borosilicate with a high crystallinity was obtained at OCTMAOH/Si ratio of 0.3, which was much smaller than the ratio required for HMI or PI. The appropriate Si/B ratio for a pure MWW borosilicate was proved to range from 10 to 20, while a structure similar to MCM-56 would emerge when the Si/B ratio was out of that range. However, a higher B content would favor the synthesis of MWW titanosilicate [41], since the introduction of Ti atoms severely hinder the crystallization process. More Ti atoms can be introduced into the MWW zeolite when the B content was maintained at a relatively high level, i.e., with a Si/B ratio of 5. The particle size of Ti-MWW zeolite synthesized using OCTMAOH as SDA was smaller than those prepared with PI and HMI molecules, which was supposed to favor the diffusion of substrate molecules and to improve the catalytic activity. Similar to above-mentioned synthesis with PI and HMI, there was a broad band in the UV–Vis spectrum of the Ti-MWW zeolite as-synthesized with OCTMAOH, indicating tetrahedral and octahedral Ti species co-existed in the obtained Ti-MWW lamellar precursor. An acid treatment and subsequent calcination can remove the octahedral and anatase Ti species selectively, with the active tetrahedral Ti species remaining in the framework of MWW. Unlike the Ti-MWW zeolite prepared with HMI or PI, more than half of the Ti atoms were removed during the post-treatment procedures. That is probably because the Ti species involved in the part of Ti-MWW zeolite with a relatively poor crystallinity can be removed easily, which again suggested that the introduction of Ti atoms largely decrease the crystallinity. With a smaller particle size, Ti-MWW prepared with OCTMAOH showed comparable activity with those prepared with PI or HMI, mainly again due to a poor crystallinity.

The effect of alkali cations as well as another two linear type ammoniums was also investigated following the above study. As shown in Table 2.3, OCTMAOH has stronger directing ability than both heptyltrimethylammonium hydroxide (HEPMAOH) and hextytrimethylammounium hydroxide (HEXAMOH), and the addition of alkali cations was helpful to the crystallization process when it was maintained at an appropriate ratio of K/Si = 0.04–0.06. Both the chemical element analysis (Table 2.4) and ^{13}CNMR analysis (Fig. 2.5) verified that the SDA molecules all well maintained their chemical structures during the crystallization process. Broadness and overlap of the resonance were observed in the ^{13}C NMR spectra of Ti-MWW zeolites compared with that of free SDA dissolved in dimethyl sulfoxide (DMSO), which was presumed to be related to the geometric constrain and van der Waals interactions with the zeolite frameworks. In the TG analysis, there were two kinds of weight loss, that is, below 473 K region and above 473 K region. The former weight loss was due to physical adsorbed water molecules while the latter was supposed to be caused by the decomposition of the organic SDAs. After an acid treatment, the weight loss above 473 K was decreased by about 50 wt% (Table 2.4), which indicated that about half of the SDAs were located in the interlayer space while the other half were remaining stuck in the intralayer 10-MR channels. These linear-type SDAs can diffuse into different channels without any constrains, not only filling the pores but also stabilizing the whole structure. Acid and subsequent calcination post-treatments were still needed to remove the extra-framework Ti species as well as the alkali cations before they were used as catalysts. However, the 101 and 102 diffractions overlapped during this post-treatment procedure (Fig. 2.6), indicating the co-existence of another configuration of MWW zeolite, MCM-56 [42, 43]. A structural disorder. was believed to occur along the layer stacking direction in MCM-56 zeolite, which possesses larger external surface than normal MWW zeolite. Hence, the Ti-MWW zeolite prepared with linear-type organic molecules showed comparable activity with that synthesized with PI molecules in spite of a poor crystallinity. The small particle sizes and the co-existence of MCM-56 were supposed to compensate the shortage aroused by the poor crystallinity of Ti-MWW prepared with linear SDA molecules.

2.3.2   Synthesis of Ti-MWW by a Dry-Gel Conversion Method

Dry-gel conversion has been proved to be a useful method to prepare several zeolites, such as MFI [44], BEA [45], FER [46], MOR [47], and MTW [48]. Compared with the traditional hydrothermal synthesis method, DGC shows many

Table 2.3 The titanosilicates synthesized at different Si/Ti and K/Si ratios by using various SDAs[a]

Organic SDA	Si/Ti = 40			Si/Ti = 50			Si/Ti = 100		
	$x \leq 0.04$	$x = 0.06$	$x \geq 0.1$	$x = 0.04$	$x = 0.06$	$x = 0.1$	$x = 0.04$	$x = 0.06$	$x = 0.1$
OCTMAOH	Amor.	MWW + Amor.	MWW + Amor.	MWW	MWW	MFI	MWW	MWW	MFI
HEPMAOH	Amor.	MWW + Amor.	MWW + Amor.	MWW	MWW	MFI	MWW	MWW	MFI
HEXMAOH	MFI	MFI	MFI	MWW	MFI	MFI	MWW	MFI	MFI

Reprinted from Ref. [41], Copyright 2011, with permission from Elsevier

[a] Gel compositions and crystallization conditions, SDA/Si = 0.3; Si/B = 10; H$_2$O/Si = 30. The crystallization was performed under static conditions at 443 K for 7 days. x represents the K/Si ratio in the gels

Table 2.4 CHN chemical analysis and weight loss of as-synthesized Ti-MWW

Sample	C/N[a]	Weight loss (wt%)[b]		
		<473 K	>473 K	Total
Ti-MWW-HEP	9.7	3.2 (1.1)	15.5 (7.8)	18.7 (8.9)
Ti-MWW-OCT	11.2	2.9 (1.3)	17.6 (10.5)	20.5 (11.8)

Reprinted from Ref. [41], Copyright 2011, with permission from Elsevier
[a] Given by chemical analyses
[b] Given by TGA. The values in parentheses show the amount after acid treatment with 2 M HNO₃

Fig. 2.5 ¹³C NMR spectra of OCTMAOH (*a*) and HEPMAOH (*c*) in DMSO-d6 solution and ¹³C CP MAS NMR spectra of Ti-MWW-OCT (*b*) and Ti-MWW-HEP (*d*) both in as-synthesized form. Reprinted from Ref. [41], Copyright 2011, with permission from Elsevier

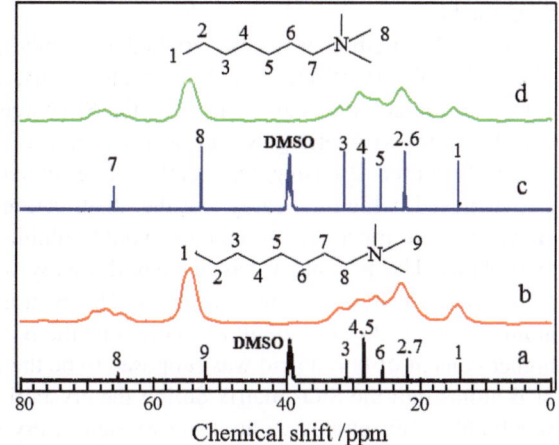

Fig. 2.6 The XRD patterns of acid-treated and calcinced Ti-MWW-OCT (Si/Ti = 50) (*a*), Ti-MWW-OCT (Si/Ti = 100) (*b*), Ti-MWW-HEP (Si/Ti = 50) (*c*), and Ti-MWW-HEP (Si/Ti = 100) (*d*). The as-synthesized samples were treated with 2 M HNO₃ at 373 K for 20 h, and then calcined at 823 K for 10 h. Reprinted from Ref. [41], Copyright 2011, with permission from Elsevier

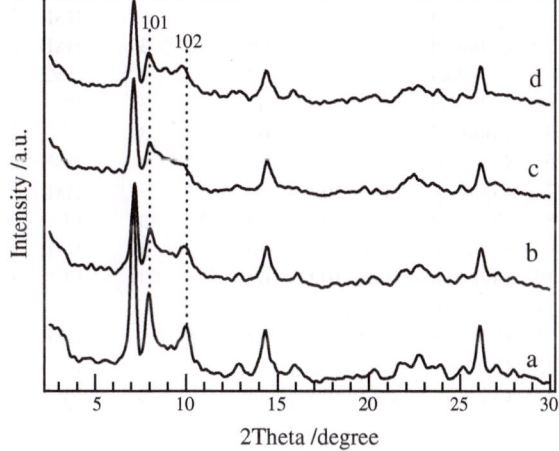

advantages, including higher yield, shorter crystallization time, and less organic material consumption. Moreover, uniform crystals with much smaller particle sizes were obtained in synthesizing both BEA-type aluminosilicate and titanosilicate with DGC method in comparison with the hydrothermal synthesis [18, 49]. The DGC method further favored the crystallization of BEA-type titanosilicate under Al-free conditions, which was much more difficult in the hydrothermal synthesis method. The DGC method can be divided into two categories in terms of the volatility of organic SDAs, that is, stream assisted crystallization (SAC) and vapor-phase transport (VPT). DGC with volatile SDAs was achieved by the VPT method, in which the SDAs and water were separated from the dried synthetic gels. As for non-volatile SDAs, they were added into the synthetic gels with water placed below them, and then the crystallization was realized with the so-called SAC method.

The DGC method was also applied to synthesis of Ti-MWW zeolite [27] (denoted as Ti-MWW-DGC) and the results are listed in Table 2.5. Both colloidal silica and fumed silica can be used as the Si source. Sodium cations were introduced into the synthetic gels with a Si/Na ratio of 37 when colloidal silica was used as Si source. By using alkali cations-free fumed silica, the crystallization was also achieved, which indicated that the alkali cations were not always necessary. However, too much alkali cations would inhibit the crystallization, yielding amorphous. The B content was decreased greatly with Si/B = 5, compared with Si/B = 0.75 in hydrothermal synthesis. The B atoms are much smaller than Si atoms, which makes it difficult to incorporate the B atoms into the framework. The higher content of boron acid was supposed to be the driving force for the insertion of B atoms into the SiO_2 matrix during the crystallization process. The B content was highly concentrated as no water existed in dry gels, and it was probably the

Table 2.5 The results of Ti-MWW-DGC synthesized by a dry-gel conversion method

No	Silica source	Gel composition			SDA	Time (week)	Product composition		
		Si/Ti	Si/B	Si/Na			Si/Ti	Si/B	Si/Na
1	Colloidal	60	5	37	HMI	2	61	17	114
2	Colloidal	60	3	37	HMI	2	58	13	135
3	Colloidal	60	1	37	HMI	2	54	13	325
4	Colloidal	60	5	37	PI	2	66	16	124
5	Colloidal	60	3	37	PI	2	84	14	–
6	Cab-o-sil	60	5	∞	HMI	2	84	14	∞
7	Colloidal + seed (10 %)	60	12	37	HMI	2	63	18	73
8	Colloidal + seed (10 %)	60	12	37	PI	2	63	18	73
9	Colloidal + seed (10 %)	60	5	37	HMI	1	–	–	–
10	Colloidal + seed (10 %)	60	5	37	PI	1	84	14	454
11	Cab-o-sil + seed (10 %)	60	8	∞	PI	1	58	14	∞
12	Cab-o-sil + seed (10 %)	60	5	∞	HMI	1	–	–	–
13	Cab-o-sil + seed (10 %)	60	5	∞	PI	1	86	14	∞
14	Cab-o-sil + seed (10 %)	30	5	∞	PI	1	28	13	∞
15	Cab-o-sil + seed (50 %)	30	92	∞	PI	1	49	155	∞

Fig. 2.7 SEM images of Ti-MWW synthesized by hydrothermal method (**a**) and dry-gel conversion method (**b**). Reprinted from Ref. [27], Copyright 2005, with permission from Elsevier

reason why less boron acid was needed in DGC method. The B content was further decreased to Si/B = 12 when 10 wt% of deboronated MWW zeolite was added as seed, meaning that the introduction of seed favored the crystallization process. Moreover, the crystallization process was speeded up upon the addition of seed and Ti-MWW-DGC with a high crystallinity was obtained within 1 week. The Si/B ratio even increased up to 92 when 50 % seed was added into the synthetic gels, the crystallization mechanism of which was not clear. The possibility of post-synthesis via reversible structural conversion occurring on the seed crystals cannot be excluded when such a high content seed was added. This aspect will be described in next section.

Both tetrahedral and octahedral Ti species were formed during the DGC crystallization process. The octahedral Ti species again can be selectively removed with an acid treatment, leaving only a sharp band around 220 nm in the UV–visible spectra. However, the acid treatments lead to the Ti-MWW-DGC zeolite with a higher Si/Ti ratio than hydrothermal synthesized Ti-MWW zeolite, indicating that more Ti species located in the non-framework sites and thus unstable in acid treatment. As mentioned above, uniform crystals with smaller particle sizes were obtained in synthesizing Ti-Beta zeolite with the DGC method. Unfortunately, the particle size of Ti-MWW-DGC was 10–20 times larger than those obtained by hydrothermal synthesis (Fig. 2.7), which certainly imposed significant diffusion problems for both the substrates and the products. Thus, Ti-MWW-DGC showed a much lower activity in the epoxidation of 1-hexene than hydrothermally synthesized Ti-MWW.

2.3.3 Synthesis of Ti-MWW with the Assistant of F^- Ions

The F^- ion is an important accessory ingredient for zeolite crystallization process. With the help of F^- ions, Ti-Beta zeolite can be synthesized under Al-free

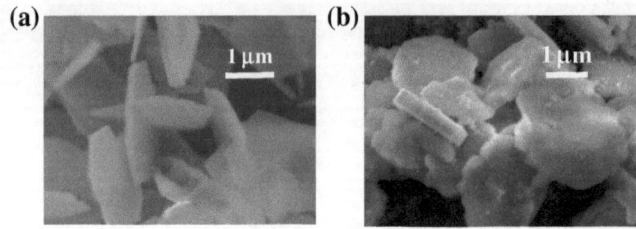

Fig. 2.8 SEM images of Ti-MWW synthesized with the assistant of F$^-$ ions using HMI (**a**) and PI (**b**) as SDA

conditions [50]. In addition, F$^-$ ions can stabilize the double 4-MR units in constructing those zeolites with extra-large pore channels [51–53], especially gemanosilicates. To decrease the amount of B atoms, the F$^-$ ions were also introduced into the synthetic gels of Ti-MWW zeolite. With the help of F$^-$ ions, Ti-MWW zeolite with a high crystallinity (denoted as Ti-MWW-F) can be obtained with the Si/B ratio varying in the range of 1–15. Similar with other zeolites produced from the F$^-$ medium[54], Ti-MWW-F possessed very large particle size and was ten times larger than Ti-MWW-HTS. In conventional hydrothermal synthesis method, Ti-MWW zeolite synthesized with HMI SDA had larger particle size than that synthesized with PI molecules, which was the opposite case in F$^-$ medium. Thus, the Ti-MWW-F-PI particle size was two times larger than that of Ti-MWW-F-HMI (Fig. 2.8). Another special phenomenon was that a band around 330 nm assigned to anatase TiO$_2$ showed up when PI was used as the SDA molecules in F$^-$ medium. When HMI was used as SDA, only tetrahedral and octahedral coordinated Ti species were involved in the Ti-MWW structure. Since inactive anatase TiO$_2$ cannot be removed by acid treatment, the HMI molecules were selected as SDA when synthesizing Ti-MWW zeolite in F$^-$ medium.

Some of the Ti-MWW-F lamellar precursor formed 3D MWW structure while some formed an interlayer expanded structure upon the acid treatment which was meant to remove the octahedral coordinated Ti species (Table 2.6). Although the inactive octahedral coordinated Ti species were removed as verified by UV–Vis spectra, the 3D Ti-MWW-F showed very low activity in the epoxidation of cyclohexene. It was supposed that the large particle size of Ti-MWW-F brought about diffusion constrains for the substrates of cyclohexane. However, another interlayer expanded structure obtained upon acid treatment showed largely enhanced activity than the 3D Ti-MWW-F zeolite. This interlayer-expanded Ti-MWW structure has been reported as Ti-YUN-1 [55]. Figure 2.9 shows the TEM images of both 3D Ti-MWW zeolite and interlayer-expanded Ti-MWW zeolite. It can be obviously observed that the layer spacing of interlayer-expanded Ti-MWW zeolite was larger than 3D Ti-MWW zeolite by 2–3 Å. And the structure was supposed to be formed by inserting monomeric Si species into the interlayer space

Table 2.6 Synthesis composition and oxidation of cyclohexene with H_2O_2

Sample No.	Gel				Acid		Cyclohexene	H_2O_2 (%)	
	Si/Ti	Si/B	HMI/Si	HF/Si	Treatment	Phase	Conversion (%)	Conversion	selectivity
1	30	6	1.4	1	50 mL:1 g	Expanded 3D	20.8	28.4	73.2
2	30	6	1.4	0.84	50 mL:1 g	Expanded 3D	13.9	13.9	99.7
3	30	6	1.4	0.6	50 mL:1 g	3D MWW	0.53	0.79	66.9
4	40	6	1	1	30 mL:1 g	3D MWW	2.9	3.3	89.1
5	50	6	1	1	30 mL:1 g	3D MWW	2.1	2.7	77.8
6	60	6	1	1	30 mL:1 g	Expanded 3D	12.3	14.1	87.2
7	70	6	1	1	30 mL:1 g	Expanded 3D	9.5	15.5	62.3
8	80	6	1	1	30 mL:1 g	Expanded 3D	10.5	14.8	70.9
9	100	6	1	1	30 mL:1 g	Expanded 3D	8.3	10.7	77.6

Conditions: cyclohexene, 10 mmol; H_2O_2 (31 wt%), 10 mmol; acetonitrile, 10 mL; cat., 50 mg; temp., 60 °C; time, 2 h

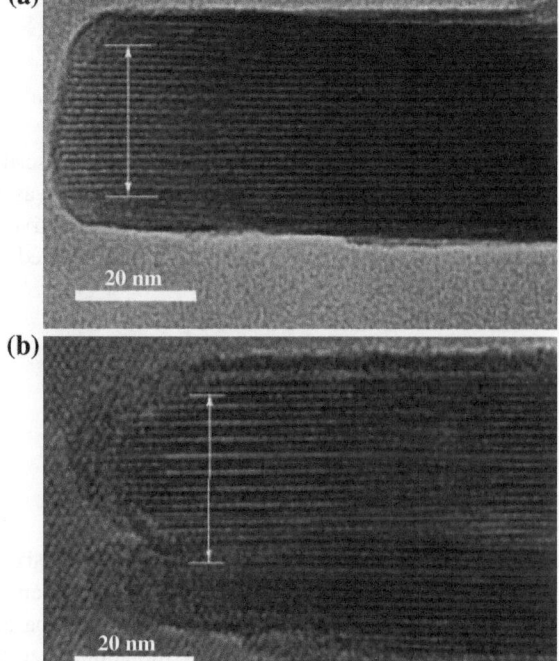

Fig. 2.9 Edge-on TEM image views of 3D Ti-MWW (**a**) and expanded 3D Ti-MWW (**b**). The *arrow* indicates the thickness of 10 MWW layers. Reprinted from Ref. [55], Copyright 2004, with permission from John Wiley and Sons

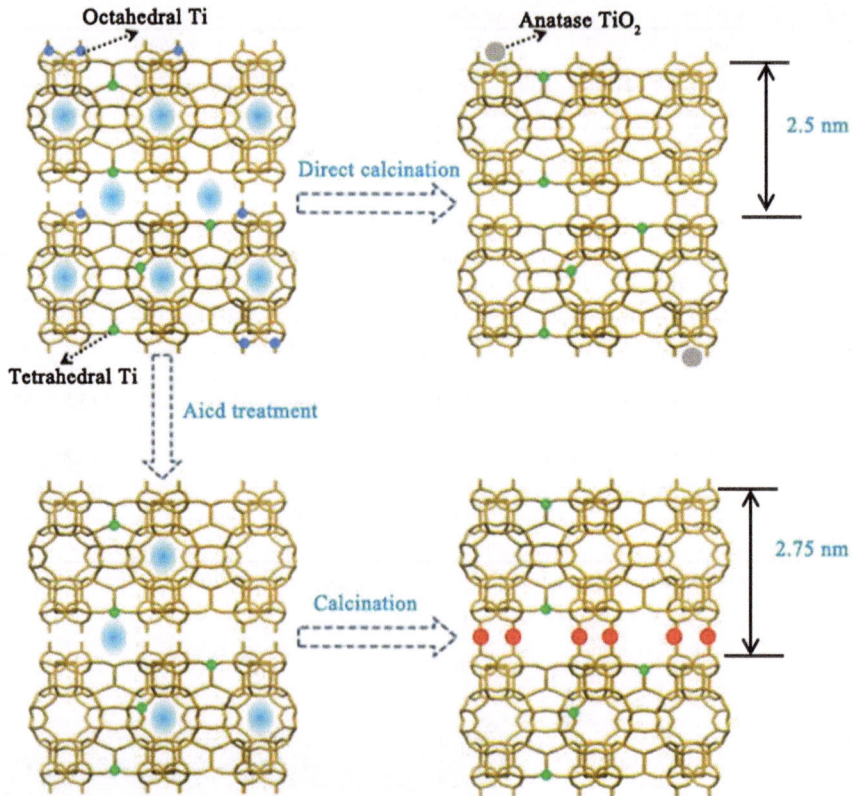

Fig. 2.10 Possible scheme for the formation of expanded Ti-MWW structure

upon acid treatment (Fig. 2.10), which will be described in detail in Sect. 3.5. With larger interlayer space, the diffusion constrain was largely released and the substrate of cyclohexene can easily reach the active Ti sites. Thus, Ti-MWW-F zeolites with interlayer expanded structure showed extremely high activities.

2.4 Boron-Free Ti-MWW Synthesis

2.4.1 Synthesis of Ti-MWW with Dual Structure-Directing Agents

Organic SDAs play two different roles in synthesizing MWW lamellar precursor, including filling the intralayer 10-MR channels and stabilizing interlayer spaces. ITQ-1 [56], pure silica MWW zeolite, was prepared with dual SDAs, HMI and TMAadOH. TMAadOH with larger molecular size locate in the interlayer spaces

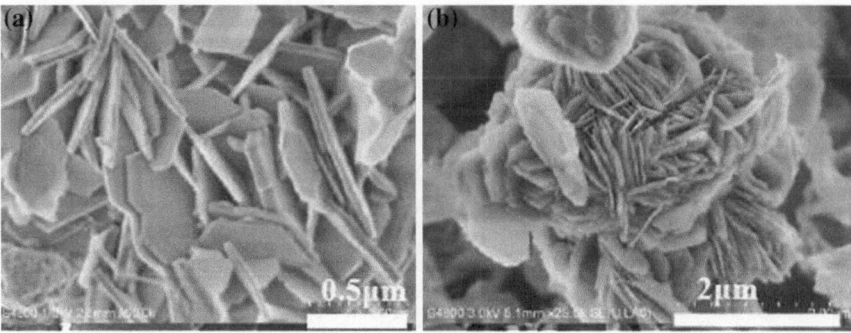

Fig. 2.11 SEM images of Ti-MWW using dual structure-directing agents with Si/Ti ratio of 50 (**a**) and 100 (**b**)

Table 2.7 Chemical composition of Ti-MWW synthesized with hydrothermal method

SDAs	%N	%C	%H	C/N	TG[a]
HMI[b]	3.03	16.45	3.41	5.82	24.6
TMAdaOH + HMI	1.81	13.30	2.77	11.49	21.1
TMAdaOH + HMI[c]	1.20	7.82	2.07	7.60	13.8
TMAdaOH[b]	1.01	10.53	1.92	12.16	16.7

[a] Weight loss by thermogravimetric analysis to 1073 K
[b] Synthesized with Si/B = 0.75
[c] Acid treatment with 2 M HNO_3

while smaller HMI molecules diffuse into the intralayer 10-MR channels. The structure of ITQ-1 zeolite was constructed by the corporation of these two organic molecules in the synthetic gels. Stimulated by this crystallization process, Liu et al. synthesized Ti-MWW zeolite (denoted as Ti-MWW-Dual) under B-free conditions by hydrothermal synthesis method with the help of the above two organic molecules and K^+ ions [57]. In the conventional hydrothermal method, a large amount of boron acid (Si/B = 0.75) was needed with either HMI or PI as SDA. Nevertheless, the B content can be decreased to zero in the dual SDAs system. On the other hand, the amount of organic SDA was decreased a lot, too. When PI or HMI was used to direct the construction of MWW structure, the ratio of PI/Si or HMI/Si was about 1.4, while the ratio in the final product was only 0.1. That is, most of the organic molecules were wasted during the crystallization process. The total amount of organic molecules in the dual structure-directing system was enough at the (HMI + TMAadOH)/Si ratio of 0.56, which was less than half of that used in the system with single SDA. Ti-MWW zeolite with a high crystallinity can be obtained when the ratio of TMAadOH/HMI ranged from 0.2 to 1.2. The product with the C/N ratio of 11.49 (Table 2.7), a value between the C/N ratio of TMAadOH molecule and HMI molecule, contained about 20 % HMI molecules and 80 % TMAadOH molecules. After an acid treatment, the C/N ratio was

Fig. 2.12 UV–visible
spectra of as-made Ti-MWW
synthesized using dual
structure-directing agents
with the Si/Ti ratio of (*a*) 100,
(*b*) 50, (*c*) 30 and (*d*) 20

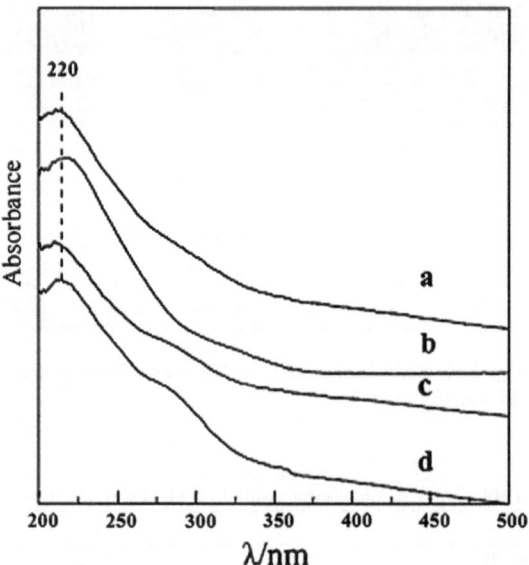

decreased to 7.60, more close to the C/N ratio of HMI molecules, meaning that most of the TMAadOH molecules were removed during the acid treatment and the remaining HMI molecules were hardly washed out of the structure. Hence, it can be inferred that HMI molecules was stuck in the intralayer 10-MR channels while the TMAadOH molecules located in the interlayer spaces, which was similar with that in synthesizing ITQ-1 zeolite.

Although the alkali cations in the synthetic gels of titanosilicates would retard the insertion of Ti atoms into the framework, minute alkali cations was necessary in the crystallization process in some special cases, such as the synthesis of Ti-MWW-Dual zeolite. When the ratio of K⁺/Si was lower than 0.05, only amorphous was obtained. The appropriate ratio of K⁺/Si was set at 0.07 after detailed investigation. Too much K⁺ cations would cause the formation of other lamellar structures other than MWW zeolite. The K⁺ cations would be removed by an acid treatment before Ti-MWW zeolite was used as the catalyst in epoxidation reactions.

As shown in the SEM images (Fig. 2.11), these crystals of Ti-MWW-Dual zeolite possessed uniform particle size of $0.5 \times 0.5 \times 0.1$ μm, which was similar with that of Ti-MWW crystals synthesized with the help of PI by hydrothermal method. The change of Ti content hardly affects the aggregation and the particle sizes. A synthetic gel with the Si/Ti ratio of 20 can still yield pure Ti-MWW zeolite, the crystallinity of which was very high, though. Only very weak band was observed around 260 nm in the UV–visible spectra (Fig. 2.12) meaning that Ti-MWW-Dou zeolite has much less octahedral Ti species than those synthesized with other methods. That is probably because Ti species can be inserted into the framework more easily without the competition of B atoms in the synthesis gels.

Fig. 2.13 FTIR spectra of
Ti-MWW synthesized using
dual structure-directing
agents (*a*) and PI (*b*) with Si/
Ti = 50 in hydroxyl
stretching region

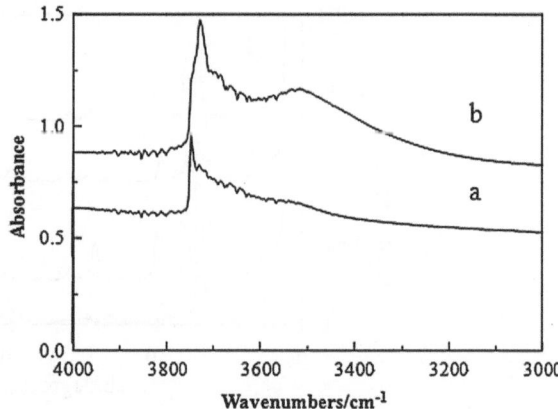

The existence of framework Ti species induced a band around 960 cm^{-1} in the IR spectra measured under air condition. The band shifted to 930 cm^{-1} when the sample was pretreated in vacuum. This was supposed to be related to the absence of B atoms, because a shoulder band would also appear besides the original band at 960 cm^{-1} in the IR spectra of Ti-MWW zeolite synthesized with B atoms pretreated in vacuum. However, the real reason is not clear. In the hydroxyl stretching vibration regions, Ti-MWW-Dou zeolite synthesized in the absence of B atoms showed a band round 3750 cm^{-1} with lower intensity than that synthesized with the help of B atoms (Fig. 2.13). The band around 3750 cm^{-1} appearing in the IR spectra of Ti-MWW samples after evacuation was attributed to the terminal Si–OH groups. B atoms with small cation radii were loosely incorporated into the framework of Ti-MWW zeolite probably inducing much more defects and more Si–OH groups. Hence, the Ti-MWW zeolite synthesized in the presence of B atoms was more hydrophilic, which was caused by the existence of numerous terminal hydroxyl groups.

With the presence of K$^+$ cations, Ti-MWW zeolite was almost inactive in the epoxidation reactions. After an acid treatment, the K$^+$ cations would be almost completely removed and Ti-MWW-Dual showed good performance as the catalysts in the epoxidation reactions, which means the existence of K$^+$ cations indeed poisoned the Ti active sites. In the epoxidation of 1-hexene, the Ti-MWW-Dou zeolite showed higher activity than that synthesized with sole organic agent, although the particle sizes of the former was a little bit larger. The absence of B atoms favored the insertion of Ti atoms, decreased the amount of defects and enhanced the hydrophobicity, which is the reason why Ti-MWW-Dual synthesized with the coexistence of HMI and TMAadOH molecules was more active. Ti-MWW-Dual zeolite was also stable enough to resist the leaching of Ti species during liquid phase reactions, the activity of which was remained even after fifth catalytic runs.

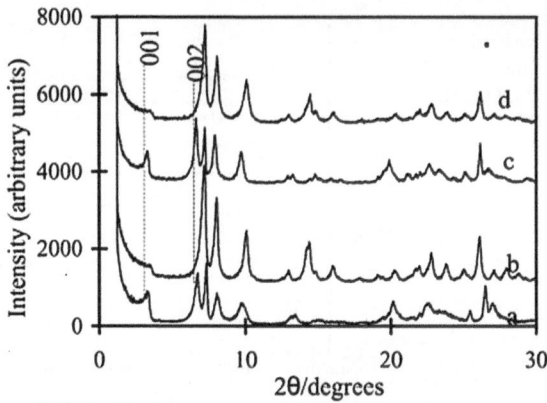

Fig. 2.14 X-Ray diffraction patterns of as-synthesized B-MWW (Si/B $= 11$) (a), deboronated MWW (Si/B > 1000) (b), Ti-MWW (Si/Ti $= 30$) precursor (c), and sample c treated with 2 M HNO$_3$ and calcined (d). Reprinted from Ref. [28], Copyright 2002, with permission from Royal Society of Chemistry

2.4.2 Synthesis of B-Free Ti-MWW Through Reversible Structure Conversion

The lamellar zeolite can be converted to a rigid 3D zeolite upon calcination, during which condensation between silanols on the up and down layers would take place. And the reversibility of this process was supposed to hardly happen because of the strong Si–O–Si bonds connecting the layers. However, several cases involving the change from 3D zeolite to 2D layered zeolite have been found recently. Germanosilicate UTL zeolite with double 4-MR structures rich in Ge atoms easily hydrolyzes to lamellar zeolite with layers similar to FER layers in mild acid conditions, which is also believed to be a promising way to expand the family of layered zeolites [58, 59]. Another example is the reversibility between the 3D MWW structure and its lamellar precursor. That is, 3D MWW structure can be changed to MWW lamellar precursor with the help of amine molecules [28].

The reversibility can be applied to preparing MWW titanosilicate in the absence of B atoms using a post-synthesis method (denoted as Ti-MWW-PS) [28]. This method involves the preparation of high silica MWW zeolite and the incorporation of Ti atoms during the structural conversion process. The structural changes can be revealed by the XRD patterns as shown in Fig. 2.14. The 001 and 002 diffraction peaks disappeared after calcination and deboronation procedures which indicated that a rigid 3D MWW structure was formed. When the deboronated MWW zeolite (Si/B > 1000) was subjected to the structural conversion process in the presence of Ti source (TBOT), the 001 and 002 diffraction peaks were restored, meaning the lamellar structure was again obtained. Several organic amines including HMI, PI, pyridine, and piperazine were employed to direct this conversion process. However, the reversible structural conversion only occurred in the presence of HMI and

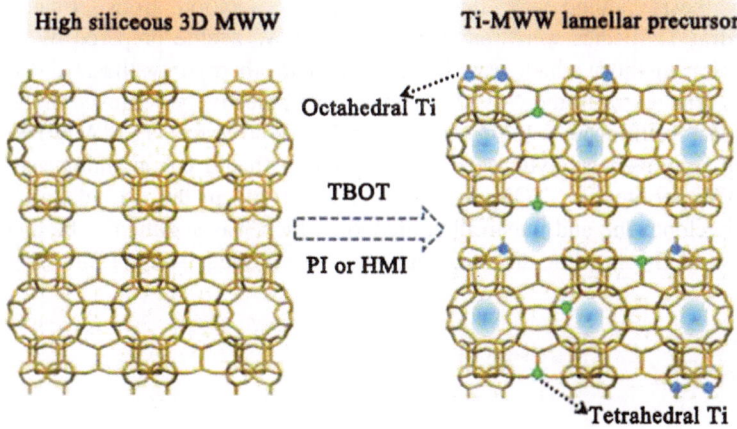

Fig. 2.15 Graphic description for the postsynthesis of boron free Ti MWW via reversible structure conversion

PI molecules, which were two typical structure-directing agents for preparing MWW zeolite. Reorganization seemed to have happened between the MWW structure and the structural-directing agents in the conversion process. The highly siliceous MWW zeolite (Si/B > 1000) was prepared through severe acid treatment of the calcined MWW borosilicate, creating numerous vacancies in the framework. The alkalinity provided by the amine molecules in the conversion process would cleave the Si–O–Si bonds between layers, making the interlayer space larger to favor the diffusion of Ti precursors into the vacancies (Fig. 2.15). All the Ti atoms in the synthetic gels can be inserted into the framework, indicating that the

Table 2.8 1-Hexene oxidation over various titanosilicates with acetonitrile as a solvent[a]

Cat	Si/Ti	1-Hexene		H_2O_2	
		Conversion %	TON	Conversion %	Efficiency %
Ti-MWW-OCTMAOH	95	16.5	176	19.2	86
	132	14.6	215	17.3	84
Ti-MWW-Dual	35	42.6	180	52.5	81
	103	42.2	488	51.2	82
Ti-MWW-HMI	46	37.5	213	40.3	93
	72	22.4	198	25.6	88
Ti-MWW-PS	58	60.7	480	67.9	93
	100	49.8	500	50.1	92
Ti-MWW-DGC	58	1.7	42	7	33
	84	2.4	69	19	16
Ti-MWW-F	70	51.1	465	58	88.1
	356[b]	16.3	706	19.7	82.7

[a] Reaction conditions: bath reactor; cat., 50 mg; substrate, 10 mmol; H_2O_2, 10 mmol; MeCN 10 mL; temp., 333 K; time, 2 h
[b] Ti-MWW with expanded structure

post-synthesis method was effective for Ti incorporation. Both tetrahedral and octahedral Ti species were found in the Ti-MWW zeolite prepared by structure conversion method, which was the same as hydrothermal synthesized Ti-MWW zeolite. The extra-framework Ti species was selectively removed by an acid treatment, resulting in a highly active catalyst. The B-free Ti-MWW-PS zeolite showed improved catalytic activity compared with the B-containing and B-less Ti-MWW zeolite, i.e., TON and epoxide selectivity in the epoxidation of both bulky cyclohexene and linear alkyl alcohol, which was assumed to be due to the decrease of B-related acidity and framework electronegativity (Table 2.8).

References

1. Bellussi G, Rigutto MS (2001) Metal ions associated to molecular sieve frameworks as catalytic sites for selective oxidation reactions. Stud Surf Sci Catal 137:911–955
2. Ratnasamy P, Srinivas D, Knözinger H (2004) Acitve sites and reactive intermediates in titanium silicate molecular sieves. Adv Catal 48:1–169
3. Wu P, Komatsu T, Yashima T (1996) Characterization of titanium species incorporated into dealuminated mordenites by means of IR spectroscopy and ^{18}O-exchange technique. J Phys Chem 100:10316–10322
4. Corma A, Camblor MA, Esteve PA et al (1994) Activity of Ti-Beta catalyst for selective oxidation of alkenes and alkanes. J Catal 145:151–158
5. Tuel A (1995) Synthesis, characterization, and catalytic properties of the new TiZSM-12 zeolite. Zeolites 15:236–242
6. Wu P, Komatsu T, Yashima T (1997) Ammoximation of Ketones over titammium mordenite. J Catal 168:400–411
7. Wu P, Komatsu T, Yashima T (1998) Hydroxylation of aromatics with hydrogen peroxide over titanosilicates with MOR and MFI structures: effect of Ti peroxo species on the diffusion and hydroxylation activity. J Phys Chem B 102:9297–9303
8. Xu H, Zhang YT, Wu HH et al (2011) Postsynthesis of mesoporous MOR-type titanosilicate and its unique catalytic properties in liquid-phase oxidations. J Catal 25:263–272
9. van der Waal JC, Rigutto MS, van Bekkum H (1998) Zeolite titanium beta as a selective catalyst in the epoxidation of bulky alkenes. Appl Catal A: General 167:331–342
10. Blasco T, Corma A, Navarro MT et al (1995) Synthesis, characterization, and catalytic activity of Ti-MCM-41 structures. J Catal 156:65–74
11. Leonowicz ME, Lawton JA, Lawton SL et al (1994) MCM-22: a molecular sieve with two independent multidimensional channel systems. Science 264:1910–1913
12. Corma A, Martínez-Soria V, Schnoeveld E (2000) Alkylation of benzene with short-chain olefins over MCM-22 zeolite: catalytic behavior and kinetic mechanism. J Catal 192:163–173
13. Corma A, Corell C, Pérez-Pariente J (1995) Synthesis and characterization of MCM-22 zeolite. Zeolites 15:2–8
14. Komura K, Murase T, Sugi Y et al (2010) Synthesis of boron-containing CDS-1 zeolite by topotactic dehydration condensation of [B]-PLS-1 prepared from layered silicate H-LDS. Chem Lett 39:948–949
15. Wu P, Liu H, Komatsu T et al (1997) Synthesis of ferrisilicate with the MCM-22 structure. Chem Commun 7:663–664
16. Ahedi RK, Kotasthane AN (1998) Synthesis of FER titanosilicates from a non-aqueous alkali-free seeded system. J Mater Chem 8:1685–1686

17. Kubota Y, Koyama Y, Yamada T et al (2008) Synthesis and catalytic performance of Ti-MCM-68 for effective oxidation reactions. Chem Commun 46:6224–6226
18. Tatsumi T, Jappar N (1998) Properties of Ti-Beta zeolites synthesized by dry-gel conversion and hydrothermal methods. J Phys Chem B 102:7126–7131
19. Lv AL, Xu H, Wu HH et al (2011) Hydrothermal synthesis of high-silica mordenite by dual-templating method. Micropor Mesopor Mater 145:80–86
20. Maschmeyer T, Ray F, Sankar G et al (1995) Heterogeneous catalysts obtained by grafting metallocene complexes onto mesoporous silica. Nature 378:159–162
21. Morey MS, O'Brien S, Schwarz S et al (2000) Hydrothermal and postsynthesis surface modification of cubic, MCM-48, and ultralarge pore SBA-15 mesoporous silica with titanium. Chem Mater 12:898–911
22. Jappar N, Xia Q, Tatsumi T (1998) Oxidation activity of Ti-Beta synthesis by a dry-gel conversion method. J Catal 180:132–141
23. Corma A, Díaz U, Fornés V et al (1999) Ti/ITQ-2, a new material highly active and selective for the epoxidation of olefins with organic hydroperoxides. Chem Commun 9:779–780
24. Levin D, Chang CD, Luo S et al (2000) US 6 114 551
25. Millini R, Perego G, Parker WO Jr et al (1995) Layered structure of ERB-1 microporous borosilicate precursor and its intercalation properties towards polar molecules. Micropor Mater 4:221–230
26. Wu P, Tatsumi T, Komatsu T et al (2001) A novel titanosilicate with MWW Structure. I. Hydrothermal synthesis, elimination of extraframework titanium, and characterizations. J Phys Chem B 105:2897–2905
27. Wu P, Miyaji T, Liu YM et al (2005) Synthesis of Ti-MWW by a dry-gel conversion method. Catal Today 99:233–240
28. Wu P, Tatsumi T (2002) Preparation of B-free Ti-MWW through reversible structural conversion. Chem Commun 10:1026–1027
29. Jorda E, Tuel A, Teissier R et al (1997) TiF4: An original and very interesting precursor to the synthesis of titanium containing silicalite-1. Zeolites 19:238–245
30. Balducci L, Bianchi D, Bortolo R et al (2003) Direct oxidation of benzene to phenol with hydrogen peroxide over a modified titanium silicalite. Angew Chem Int Ed 42:4937–4940
31. Wang X, Guo X (1999) Synthesis, characterization and catalytic properties of low cost titanium silicalite. Catal Today 51:177–186
32. Wu P, Tatsumi T, Komatsu T et al (2001) A novel titanosilicate with MWW structure: II. Catalytic properties in the selective oxidation of alkenes. J Catal 202:245–255
33. Zones SI, Santilli D (1992) In: Von Ballmoos R, Higgins JB, Treacy MMJ (eds) Proceedings of the 9th international zeolite conference, Montreal. Butterworth-Heinemann, Stoneham, p 171
34. Wagner P, Nakagawa Y, Lee GS et al (2000) Guest/host relationships in the synthesis of the novel cage-based zeolites SSZ-35, SSZ-36, and SSZ-39. J Am Chem Soc 122:263–273
35. Puppe L (1984) U.S. Pat., 4439409
36. Zones SI, Hwang S-J, Davis ME (2001) Studies of the synthesis of SSZ-25 zeolite in a "Mixed-Template" system. Chem Eur J 7:1990–2001
37. Zones SI (1989) U.S. Pat., 4826667
38. Lee S-H, Shin C-H, Hong SB (2003) Synthesis of zeolite MCM-22 using N,N,N,N′,N′,N′,-Hexamethyl-1,5-pentanediaminium and alkali metal cations as structure-directing agents. Chem Lett 32:542–543
39. Hong SB, Min H-K, Shin C-H et al (2007) Synthesis, crystal Structure, characterization, and catalytic properties of TNU-9. J Am Chem Soc 35:10870–10885
40. Liu N, Liu YM, Yue CC et al (2007) A new synthesis route for MWW analogues using octyltrimethylammonium cations as structure-directing agents under alkali-free conditions. Chem Lett 36:916–917
41. Yue CC, Xie W, Liu YM et al (2011) Hydrothermal synthesis of MWW-type analogues using linear-type quaternary alkylammonium hydroxides as structure-directing agents. Micropor Mesopor Mater 142:347–353

42. Fung AS, Lawton SL, Roth WJ (1994) U.S. Pat., 5362697
43. Corma A, Diaz U, Fornés V et al (2000) Characterization and catalytic activity of MCM-22 and MCM-56 compared with ITQ-2. J Catal 191:218–224
44. Bandyopadhyay R, Kubota Y, Sugimoto N et al (1999) Synthesis of borosilicate zeolites by the dry gel conversion method and their characterization. Micropor Mesopor Mater 32:81–91
45. Hari Prasad Rao PR, Matsukata M (1996) Dry-gel conversion technique for synthesis of zeolite BEA. Chem Commun 12:1441–1442
46. Matsukata M, Nishiyama N, Ueyama K et al (1996) Crystallization of FER and MFI zeolites by a vapor-phase transport method. Micropor Mater 7:109–117
47. Wang J, Cheng X, Guo J et al (2006) High-silica MOR type zeolite self-transformed from dry aluminosilicate gel in OSAs-free and fluoride-free reactant system. Micropor Mesopor Mater 96:307–313
48. Matsukata M, Ogura M, Osaki T et al (1999) Conversion of dry gel to microporous crystals in gas phase. Topics Catal 9:77–92
49. Hari Prasad Rao PR, Leon y Leon CA, Ueyama K et al (1998) Synthesis of BEA by dry gel conversion and its characterization. Micropor Mesopor Mater 21:305–313
50. Blasco T, Camblor MA, Corma A et al (1996) Unseeded synthesis of Al-free Ti-β zeolite in fluoride medium: a hydrophobic selective oxidation catalyst. Chem Commun 20:2367–2368
51. Blasco T, Corma A, Díaz-Cabanas MJ et al (2004) Synthesis, characterization, and framework heteroatom localization in ITQ-21. J Am Chem Soc 126:13414–13423
52. Jiang J, Jorda JL, Yu J et al (2011) Synthesis and structure determination of the hierarchical meso-microporous zeolite ITQ-43. Science 333:1131–1133
53. Hernández-Rodríguez M, Jordá JL, Rey F et al (2012) Synthesis and structure determination of a new microporous zeolite with large cavities connected by small pores. J Am Chem Soc 134:13232–13235
54. Schreyeck L, Caullet P, Mougenel J-C et al (1995) A layered microporous aluminosilicate precursor of FER-type zeolite. J Chem Soc Chem Commun 21:2187–2188
55. Fan WB, Wu P, Namba S et al (2004) A titanosilicate that is structurally analogous to an MWW-type lamellar precursor. Angew Chem 117:6877–6881
56. Camblor MA, Corell C, Corma A et al (1996) A new microporous polymorph of silica isomorphous to zeolite MCM-22. Chem Mater 8:2415–2417
57. Liu N, Liu YM, Xie W et al (2007) Hydrothermal synthesis of boron-free Ti-MWW with dual structure-directing agents. Stud Surf Sci Catal 170:464–469
58. Roth WJ, Shvets OV, Shamzhy M et al (2011) Postsynthesis transformation of three-dimensional framework into a lamellar zeolite with modifiable architecture. J Am Chem Soc 133:6130–6133
59. Verheyen E, Joos L, Van Havenbergh K et al (2012) Design of zeolite by inverse sigma transformation. Nat Mater 11:1059–1064

Chapter 3
Structural Modification of Ti-MWW: A Door to Diversity

Abstract Post-treatment converting 3D Ti-MWW zeolite to 2D Ti-MWW zeolite with the aid of ammonium molecules and subsequent calcination can largely increase the hydrophobicity of Ti-MWW zeolite and its catalytic performance. Moreover, the Ti-MWW lamellar precursor with weak hydrogen bonds in the interlayer space can also be structurally modified by swelling, partial or full delamination, and pillaring. The modified structures including partial or full delaminated and pillared structures possessed larger external surface area and higher accessibility than the conventional 3D Ti-MWW zeolite, which then showed higher conversion rate in the liquid oxidation reactions, especially, in the epoxidation of large-size cyclohexene molecule.

Keywords Ti-MWW · Structural conversion · Swelling · Partial or full delamination · Pillaring · Silylation

3.1 Introduction

In the big family of titanosilicates, Ti-MWW zeolite was a remarkable one because of its special lamellar structure. Among the 206 kinds of zeolites [1], more than 20 kinds of zeolites have their corresponding lamellar precursors, forming an important family. As a special family, the lamellar zeolites show different characteristics from the conventional 3D zeolites with rigid structures. The connection between the layers of lamellar zeolites is hydrogen bonds, which are much weaker than the chemical bonds in 3D zeolites, and that is the reason why lamellar zeolites can be modified. The layered zeolite was first obtained in the hydrothermal synthesis process of 3D zeolites. However, the reason why chemical connection between layers can be disturbed by such synthesis gel has not been identified yet. Most recently, two new technologies including both direct synthesis and post-treatment have been applied to obtain novel layered zeolites. The first is the successful synthesis of multilayered MFI zeolite [2] with a special designed

P. Wu et al., *MWW-Type Titanosilicate*,
SpringerBriefs in Green Chemistry for Sustainability,
DOI: 10.1007/978-3-642-39115-6_3, © The Author(s) 2013

Gemini-type quaternary ammonium surfactant as SDA, which was composed of a quaternary ammonium head with the ability to direct the formation of MFI nanosheets and a long alkyl chain to disturb the growth of the nanosheets along the third direction. Only the *h0k* reflections are observed in XRD patterns of layered MFI zeolite, which is indicative of an orientated growth of the zeolite crystallites along the *ac* planes as well as an interrupted stacking of comprising layers. With the same design notion, layered zeolites with topology of BEA and MTW were also synthesized by adjusting the structure of quaternary ammonium head as well as the compositions of synthetic gels. The second new method was to hydrolysis UTL zeolite to lamellar zeolites with FER layers. The clever point of this new synthesis method was to make full use of the structural peculiarity of UTL zeolite as well as the instability of germanosilicates. UTL zeolite was made up of FER layers connected by double four-member ring (D4MR) structure units which are rich in germanium atoms [3, 4]. Once the UTL zeolite was subjected to water or mild acid aqueous solution, the D4MR units were all destroyed, leaving FER layers stacking along the *a* axis. Then hydrolysis product ICP-1P can be treated as lamellar zeolite for following modifications. The above new methods have extended our horizons of synthesizing layered zeolites and it is believed that more and more layered zeolites can be prepared in the near future.

The driving force for researchers to prepare new layered zeolites is their modifiable structures. The weak connections between the neighboring layers make layered zeolites modifiable to yield several analogues with either larger surface area or larger pore sizes. Layered zeolite with MWW topology was the most extensively studied one and has been served as the model for studying other layered zeolites. Most of the post-synthesis modification techniques were first carried out on aluminosilicate MWW and then expanded to titanosilicate MWW as well as other lamellar zeolites such as PREFER [5], Nu-6 [6], PLS-4 [7] and so on. Taking titanosilicate MWW as the example, all the existing modification techniques are summarized in Fig. 3.1, including swelling [8], full delamination [8], partial delamination [9], pillaring [10], and silylation [11, 12]. As can be seen in the figure, swelling treatment is to introduce surfactant molecules into the inter-layered space of lamellar zeolites yielding an intercalated structure. Full delaminated structure with only few cell units in one ultra-thin sheet can be achieved by imposing ultrasonic treatment on the swollen structure. With most of the layer surface exposed outside, delaminated structures possess much higher surface area and external surface area than their corresponding 3D structures. In contrast, partially delaminated structures obtained directly by treating lamellar precursors with very mild acid aqueous solution and following calcination also possess higher external surface area than 3D structures but better preserved layer structure than full delaminated structure. Pillaring treatment involves mixing the swollen materials with the metal alkoxide, generally tetraethyl orthosilicate (TEOS), followed by hydrolysis in water and calcination to give final pillared structures with newly formed mesopore systems in the interlayer space. Moreover, silylation treatment carried out in acid aqueous solution with the help of monomeric silanes is a kind of well-controlled pillaring process yielding a structure with enlarged pore channels

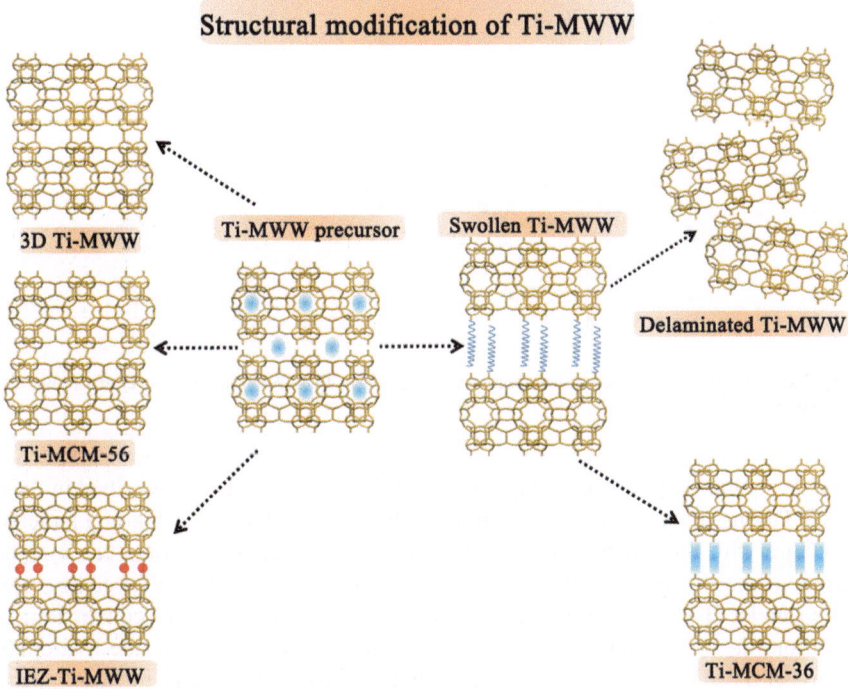

Fig. 3.1 Scheme of structural modification of Ti-MWW

between layers. All the above post-synthesis modifications favor the accessibility of substrates to active sites either by exposing the active sites outside surface or by creating opener pore system to decrease the diffusion constrains. The details about modifications on Ti-MWW zeolite will be described one by one in the following text.

3.2 Structural Conversion to Enhance the Hydrophobicity

Conversion from 2D lamellar structure to 3D rigid structure can be easily achieved by calcination, for example, lamellar zeolite PREFER to FER [13] zeolite, PLS-1 to CDS-1 [14], RUB-39 to RUB-41 [15], etc., during which topotatic condensations usually happen. However, the conversion from 3D structure to corresponding 2D structure rarely take place because of the strong Si–O–Si bonds between layers with the exception of UTL zeolite and 3D MWW zeolite. Germanosilicate UTL can be changed to layered zeolite with FER layers because the interlayer connection of Ge–O–Ge is very weak under moist or acid conditions and can be easily hydrolyzed [3, 4]. Another exception is preparing Ti-MWW through post-synthesis of 3D

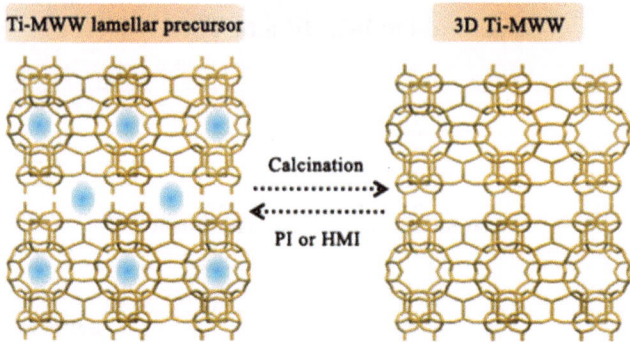

Fig. 3.2 Scheme for the structural conversion between Ti-MWW lamellar precursor and 3D Ti-MWW

MWW zeolite with Si/B ratio higher than 1000, which has been described in Sect. 2.4.2. During this process, 3D structure can be converted to corresponding lamellar structure with the aid of ammonium molecules, and titanium atoms can also be inserted into the framework simultaneously. Thus, Ti-MWW zeolite can be prepared under B-free conditions taking advantages of the structural conversion. In addition, the special structural conversion can be used to enhance the hydrophobicity of Ti-MWW zeolite resulting in higher activity especially in liquid phase reactions with H_2O_2 as the oxidant [16]. Hydrophobicity is very important to titanosilicate catalysts besides crystallinity, porosity, and surface area. Many methods have been developed to increase the hydrophobicity including post-modification of silylation [17, 18] and introducing organic groups using direct synthesis method [19, 20]. However, the organic groups introduced by silylation may block the pore channels and cause diffusion constrains. What's more, rearrangement realized usually by calcination lead to the loss of organic groups introduced either by post-synthesis or direct synthesis methods. As for Ti-MWW zeolite, enhancing hydrophobicity by structural conversion method successfully avoided the problems mentioned above.

This structural rearrangement involved treating 3D Ti-MWW zeolite with ammonium molecules under hydrothermal conditions and following calcination to yield Re-Ti-MWW zeolite with higher hydrophobicity (Fig. 3.2). Structural conversion from 3D to 2D can only be achieved by PI and HMI, with other ammonium molecules leading to either structural collapse or other zeolites, indicating the host–guest interaction between the ammonium molecules and MWW structure. This structural conversion was completed within 1 day with PI/Si ratio of 0.1. Although increasing the PI amount made the diffraction peaks of 001 and 002 more intense, the percentage of organic molecules incorporated in the sample was almost the same in despite of the changing of PI/Si ratio, indicating the PI/Si ratio of 0.1 was enough to achieve the structural conversion. Although Ti atoms in the

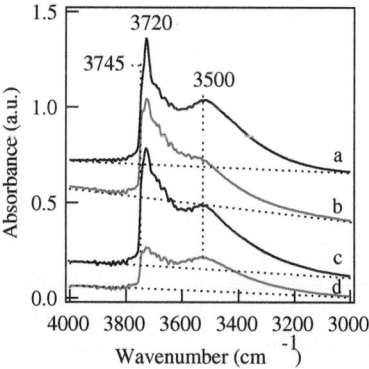

Fig. 3.3 IR spectra in the region of hydroxyl stretching vibration of the parent Ti-MWW with a Si/Ti molar ratio of 37 (*a*), its corresponding PI-treated and further calcined sample (*b*), the parent Ti-MWW with a Si/Ti molar ratio of 99 (*c*), and its corresponding PI treated and further calcined sample (*d*). Reprinted from Ref. [16], Copyright 2008, with permission from American Chemical Society

framework are more fragile than Al atoms, tetrahedrally coordinated Ti species in the framework of Ti-MWW zeolite were well preserved during this rearrangement process. The band in UV–Visible spectrum, assigned to the charge transfer from O^{2-} to Ti^{4+}, shifted from 220 to 215 nm after structural conversion, probably due to less water adsorption on the less hydrophilic Re-Ti-MWW zeolite. In contrast, B atoms suffered a loss during this rearrangement process because of their small ionic radius than Si atoms.

The IR spectra in the region of hydroxyl stretching vibration and ^{29}Si NMR spectra gave us more solid proofs for the hydrophobicity/hydrophilicity change of Ti-MWW zeolite. As shown in Fig. 3.3, these main bands at 3745, 3720, and 3500 cm^{-1} appeared in the IR spectra of Ti-MWW zeolite and Re-Ti-MWW zeolite, attributed to terminal silanols on the crystal surface, asymmetric hydrogen-bonded silanols and silanols nests [21], respectively. The internal silanols probably originated from the defects as a result of deboronation and incomplete condensation upon calcination. After the rearrangement process, the band at 3720 and 3500 cm^{-1} related to the internal silanols suffered a great decrease. It can be calculated that 35–37 % of the hydroxyl groups can be repaired during this rearrangement process. The ^{29}Si NMR spectrum (Fig. 3.4) showed that Ti-MWW zeolite contained two kinds of resonance bands originated from Q^4 groups in the region of −105 to −130 ppm and Q^3 groups at −103 ppm [22, 23]. As for Re-Ti-MWW zeolite, it showed similar ^{29}Si NMR spectrum with resonance bands of Q^4 groups but less intense resonance band of Q^3 groups, which was in accordance with the change in IR spectra. The decrease of Q^3 groups was due to the decrease of framework defects during the rearrangement process.

The decrease of silanols or framework defects was believed to enhance the hydrophobicity of titanosilicate MWW and then resulted in higher activity in

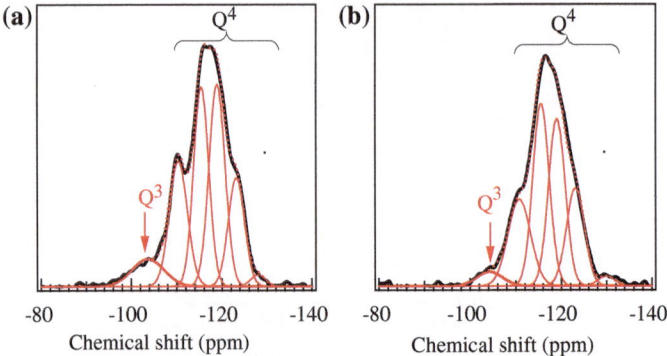

Fig. 3.4 ^{29}Si MAS NMR spectra of the parent Ti-MWW (Si/Ti = 37) (*a*) and PI-treated and further calcined sample (*b*). Reprinted from Ref. [16], Copyright 2008, with permission from American Chemical Society

liquid phase reactions. At first, an experiment to measure the amount of absorbed water of pre-dehydrated sample was carried out to identify the relative hydrophobicity of Re-Ti-MWW. Both Ti-MWW and Re-Ti-MWW zeolite were calcined at 773 K to remove absorbed water and then equilibrated with the water supplied by the saturated NH$_4$Cl solution. TG analysis showed that the amount water absorbed by Re-Ti-MWW zeolite was 81–85 % of Ti-MWW zeolite, indicating the hydrophobicity of MWW-type titanosilicate was enhanced due to the removal of silanols by rearrangement process. Then, both Ti-MWW and Re-Ti-MWW zeolite were used to catalyze the liquid-phase ammoximation of ketones and epoxidation of various alkenes to check whether the enhanced hydrophobicity would favor the catalytic performance. With comparable Ti content and the same weight ratio of catalyst to substrate, the cyclohexanone and methyl ethyl ketone conversion over Re-Ti-MWW was enhanced by 6.6 and 9 %, respectively, compared with the original Ti-MWW zeolite. As for the epoxidation of 1-hexene and propylene, 1-hexene conversion and propylene yield increased with the increase of Ti content on both Ti-MWW and Re-Ti-MWW, but the latter showed higher activity by 10–20 %. What's more, Re-Ti-MWW zeolite also showed higher activity than Ti-MWW in the epoxidation of bulk alkene molecules, including cyclohexene, cycloheptene, and cyclooctene. Thus, it can be concluded that enhanced hydrophobicity due to partial removal of silanols in defect sites indeed leads to better performance of Re-Ti-MWW zeolite.

During the rearrangement process, the migration of Si species into neighboring hydroxyl nests or dehydration of the vicinal silanols to form Si–O–Si linkage was believed to have happened to mend the disconnection and increase the hydrophobicity. So this rearrangement process with the help of amine molecules has little influence on framework Ti species but repair the MWW structure leading to a more rigid and active catalyst.

3.3 Swelling and Full Delamination of Ti-MWW

The most interesting subject in the area of micropore materials at present is to decrease the diffusion constrain because the pore size of micropore zeolite are smaller than 1 nm, which is comparable to the sizes of substrate molecules. Two general methods including exposing as many active sites outside crystal surface as possible and creating new pore channels with opener entrance have been used to solve the diffusion problems. Preparing zeolite crystals with much smaller particle sizes in the nanometer scale is one of the methods to obtain a material with larger external surface area, which would also favor the distribution of active sites on the outer surface [24]. In addition, delamination is a general way to increase the external surface area of layered zeolites [25]. A pre-swelling process and the following ultrasonic treatment were always combined to give a delaminated structure. Surfactant molecules with a long alkaline tail were introduced into the interlayer space of layered zeolite at first. Under alkaline conditions, the original hydrogen bonding between the as-synthesized lamellar zeolites is broken and the silanols are deprotonated to form SiO^- moieties, the repulsion between which would make interlayer space larger to favor the intercalation of surfactant molecules. The alkaline condition is usually supplied by tetrapropylammounim hydroxide (TPAOH), which is carefully selected to avoid competing with the surfactant molecules to enter into the interlayer space and balance the SiO^- groups [25]. An expanded structure is obtained in the pre-swelling treatment and then an ultrasonic treatment is carried out to give the final delaminated structure with largely enhanced surface area. Delamination is a versatile method and has got great success over many zeolites, such as titanosilicate PREFER [5], aluminosilicate Nu-6 [6] and so on. Titanium containing delaminated MWW zeolite was also obtained via post-synthesis method. Pure silicate MWW was at first delaminated to ITQ-2 and then titanium species were grafted to the abundant silanols exposed outside the surface [26]. However, Ti/ITQ-2 acted more like titanium containing mesoporous material, with superior performance only under anhydrous conditions. Moreover, Ti/ITQ-2 suffered serious Ti leaching and structural degradation during the liquid phase reactions.

Since the synthesis technology of Ti-MWW zeolite has been well developed, delamination can be directly realized over Ti-MWW zeolite to yield a titanium containing catalyst with high activity. Figure 3.5 shows a typical procedure for preparing delaminated materials, Del-Ti-MWW. To get active catalysts, octahedrally coordinated Ti species should be removed from the as-synthesized Ti-MWW zeolite by acid treatment before it was subjected to swelling process. 001 and 002 diffraction peaks were almost disappeared after the acid treatment because most of the organic SDA molecules intercalated in the interlayer space were removed, which decreased the interlayer space from 2.7 to 2.63 nm [8]. Nevertheless, acid treatment only caused partial destruction of Ti-MWW zeolite while the following calcination would lead to a complete condensation of the layers giving a 3D Ti-MWW zeolite (Fig. 3.5b). Swelling process was performed over

Fig. 3.5 XRD patterns: as-synthesized Ti-MWW (*a*), calcined Ti-MWW (*b*), swollen Ti-MWW with CTMABr and TPAOH at 353 K (*c*), and delaminated Ti-MWW, Del-Ti-MWW after calcination (*d*)

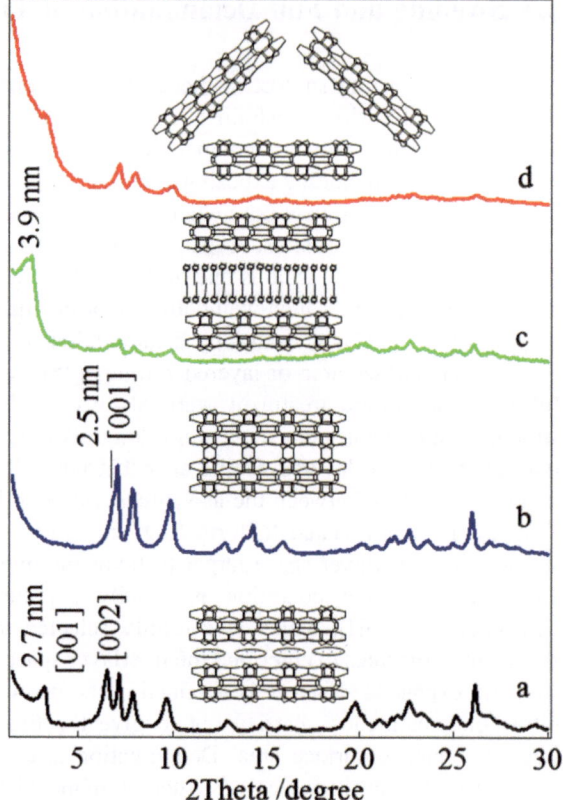

the acid treated sample in an alkaline solution containing TPAOH and surfactant hexadecyltrimethylammonium bromide (CTAB). The swollen Ti-MWW zeolite (Fig. 3.5c) showed a diffraction peak at lower 2Θ with the corresponding *d* space of 3.9 nm indicating an expanded structure was formed as a result of the intercalation of surfactant molecules. After ultrasonification and calcination, the mesophase-related structure collapsed and a delaminated material with less intensive diffraction peaks was obtained (Fig. 3.5d).

The use of TPAOH was meant to cleave the connection between the neighboring layers. With the increase of TPAOH, the diffraction peaks due to the expanded structure became more intense while those due to the MWW layers disappeared gradually indicating more disorder structure was obtained. The surface area and outer surface area also increased with the increase of TPAOH. However, every coin has two sides. Desilication was the back side, which was revealed by the weight loss after swelling. As a result, the amount of TPAOH should be well controlled to successfully cleave the interlayer connection without dissolving the MWW layers too much. The optimal amount of TPAOH was decided by the performance of delaminated catalysts used in the epoxidation of cyclopentene and cyclododecene. And the highest TON value was obtained when the weight of TPAOH to the acid-treated Ti-MWW was set at 6.0. Then the effect of ultrasonic treatment and acidification was also considered. The swollen Ti-MWW zeolite together with its swelling mixture was further ultrasonically treated and acidified with concentrated nitric acid to pH < 2, which lead to a further enhancement of the surface area of the Ti-MWW and activity in the epoxidation reactions. The ultrasonic treatment in acidic conditions was supposed to force apart MWW layers that has been intercalated by the organic surfactant molecules. It can be concluded that a profound delamination was caused by the combination of swelling and delamination.

Compared with 3D Ti-MWW zeolite, Del-Ti-MWW exhibited a great enhanced band at 3742 cm^{-1} (Fig. 3.6a) in the IR spectra, which was attributed to the isolated silanols on the layer surface. Those silanol groups were suggested to form when the interlayer connection was broken by the swelling solution. 3D Ti-MWW zeolite showed a weak band around 960 cm^{-1} assigned to the vibration of framework Ti–O–Si (Fig. 3.6b). Since the UV–visible spectra of Del-Ti-MWW has proved that tetrahedrally coordinated Ti species were hardly affected in the delamination process, Del-Ti-MWW also showed an IR band around 960 cm^{-1} but with much higher intensity, which was caused by the existence of abundant silanols on the outer surface of Del-Ti-MWW zeolite. HRTEM images were applied to further elucidate the structure of Del-Ti-MWW. As shown in Fig. 3.7, the images were taken along the *ab* plane. The crystals of Del-Ti-MWW contained many fewer sheets than 3D-Ti-MWW and even single sheet can be found in Del-Ti-MWW material, which is in agreement with the extremely large surface area of Del-Ti-MWW.

The catalytic performance of Del-Ti-MWW was compared with other titanosilicates in the epoxidation of alkenes with different sizes (Table 3.1). From linear alkenes to cyclic alkenes, the TON (turnover number) value decreased

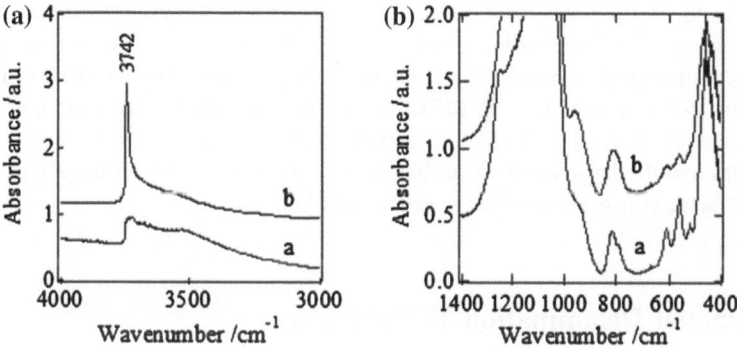

Fig. 3.6 IR spectra in the region of hydroxyl stretching vibration after the evacuation at 773 K (**A**) and IR spectra in the framework vibration region taken with KBr techniques (**B**). The samples correspond to 3D Ti-MWW (*a*) and delaminated Ti-MWW (*b*). Reprinted from Ref. [8], Copyright 2004, with permission from American Chemical Society

Fig. 3.7 HRTEM images of single sheet (**a**) and three sheets (**b**) in Delaminated Ti-MWW

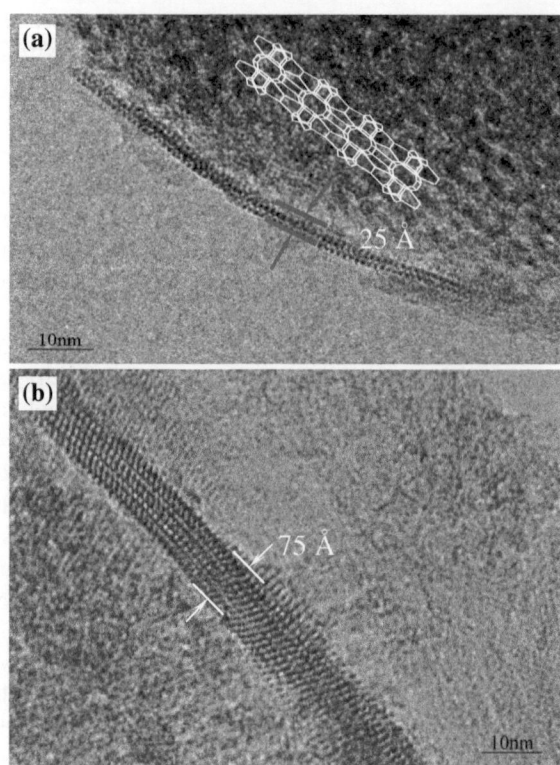

sharply for TS-1, Ti-Beta and 3D-Ti-MWW zeolites. In contrary, mesoporous catalyst Ti-MCM-41 showed very high activity in the epoxidation of cyclic alkenes while it was nearly inactive in catalyzing linear alkenes, because of the poor crystallinity of pore walls and hydrophobic nature. Del-Ti-MWW zeolite showed higher TON value than 3D-Ti-MWW zeolite for both the linear and cyclic alkenes. As for the bulk substrates, the epoxidation was supposed to happen on the outer surface. The increased external surface area caused by delamination favored the accessibility of substrate to active sites. That is why Del-Ti-MWW showed superior performance than Ti-MCM-41 in the epoxidation of bulk substrates. Delamination has made Ti-MWW zeolite possess both the crystallinity of micropores and the large external surface area of mesopores, which largely decrease the diffusion constrains and increase the catalytic activity.

3.4 Partial Delamination of Ti-MWW

Del-Ti-MWW, a delaminated derivate of titanosilicate MWW described above, possesses extremely high external surface area and superior activity in liquid phase oxidation reactions. However, the whole delamination process is somewhat

Table 3.1 Alkene epoxidation with H_2O_2 over various titanosilicates[a]

Catalyst	Si/ Ti	Surface area/ $m^2 g^{-1}$	Alkene epoxidation[b]									
			1-hexane		2-hexenes		Cyclopentene		Cyclooctene		Cyclododecenes	
			Conversion	TON	Conversion	TON (trans/cis = 61/39)	Conversion	TON	Conversion	TON	Conversion (cis/trans = 70/30)	TON
Del-Ti-MWW1[c]	42	1075	51.8	1390	89.5	2352 (81/19)	58.9	306	28.2	147	20.7	57 (53/47)
Del-Ti-MWW2[d]	39	810	24.8	863	53.7	1305 (84/16)	34.7	163	14.4	76	16.4	40 (55/45)
3D Ti-MWW	46	520	29.3	934	40.5	1053 (83/17)	15.7	89	4.3	24	3.3	9 (68/32)
TS-1	34	525	12.0	49	24.5	105 (36/64)	16.3	69	1.6	7	1.2	3 (50/50)
Ti-Beta	35	621	6.0	26	9.2	40 (35/65)	9.9	43	4.6	20	1.9	4 (56/44)
Ti-MCM-41	46	1144	0.5	3	2.3	13 (67/33)	3.5	20	5.1	29	4.1	12 (63/37)

Reprinted from Ref. [8], Copyright 2004, with permission from American Chemical Society

[a] Reprinted from Ref. [8], Copyright 2004, with permission from American Chemical Society

[a] Reaction conditions: cat., 10–25 mg; alkene, 2.5–10 mmol; H_2O_2, equal to the alkene amount; CH_3CN, 5–10 mL; temp., 313 K for cyclopentene and 333 K for other substrates; time, 2 h

[b] Conversion in mol%; TON in mol (mol Ti)$^{-1}$ which was calculated by dividing the amount of alkene converted with that of Ti amount used

[c] Delaminated with ultrasonic treatment and pH adjustment

[d] Delaminated without ultrasonic treatment and pH adjustment

troublesome. In addition, the alkaline aqueous solution adopted in the pre-swelling process lead to a deconstruction of MWW layers in some measure. To overcome the above shortcomings, a partial delaminated material analogous to MCM-56 then was proposed [9, 27], and prepared by post-synthesis method over Ti-MWW lamellar precursor. MCM-56 used to be prepared by hydrothermal method with the same synthesis gels as MCM-22 but shorter crystal time, which cannot be well controlled [28, 29].The post-synthesis method is very convenient including a mild acid treatment at lower temperature and following calcination.

Ti-MWW lamellar precursor contained well resolved 001, 002, 100, 101, 102, and 310 diffraction peaks. When a mild acid treatment was carried on the Ti-MWW, those diffraction peaks related to c axis including 001 and 002 disappeared while 101 and 102 peaks suffered a serious overlap and formed a broad diffraction peak, indicating that the arrangement along c axis was disturbed. The unchanged diffraction peaks of 100 and 310 revealed that MWW layers were well preserved during the mild acid treatment. The newly formed broad diffraction peak can be taken as a character of Ti-MCM-56 structure. It is suggested that the mild acid treatment removed part of the SDA molecules in the interlayer space, leading to disordered stacking style along the c axis.

The temperature of this mild acid treatment should be well controlled. As can be seen in Table 3.2, only 3D Ti-MWW zeolite can be obtained when a refluxing acid treatment was carried out. And it has been investigated that the temperature for acid treatment should be controlled lower than 353 K. Moreover, the size of the crystal particles also plays an important role in post-synthesis of Ti-MCM-56. Ti-MWW precursor can be hydrothermally prepared with either PI or HMI molecules as the SDA. The particle size of Ti-MWW-HMI is much larger than that of

Table 3.2 A comparison of different acid treatments on various MWW lamellar precursors of metallosilicates

Name	MWW type lamellar precursors				Products by post-treatment[a]			
	SDA	Si/M[b]	Crystal size[c]/μm		Acid treatment at 303 K		Acid refluxing	
			Thickness	Length	Si/M[b]	Structure[d]	Si/M[b]	Structure[d]
Ti-MWW-PI	PI	21	0.05–0.1	0.2–0.5	36	MCM-56	37	3D MWW
Ti-MWW-HMI	HMI	32	0.1–0.2	1–1.5	45	3D MWW	62	3D MWW
Ti-MWW-PI-F[e]	PI	23	0.4–0.5	1.5–2	32	3D MWW	37	3D MWW
B-MWW	PI	11	0.05–0.1	0.2–0.5	18	MCM-56	35	3D MWW
Al-MWW	HMI	14	0.05–0.1	0.3–0.5	16	MCM-56	19	3D MWW
Ga-MWW	HMI	15	0.05–0.1	0.3–0.5	17	MCM-56	–	–
Fe-MWW	HMI	16	0.05–0.1	0.2–0.3	20	MCM-56	25	3D MWW

Reprinted from Ref. [9], Copyright 2008, with permission from Elsevier

[a] The acid treatment was carried out on the MWW lamellar precursor in 2 M HNO_3 for 18 h followed by further calcination at 803 K for 6 h

[b] The silicon to metal molar ratio, determined by ICP

[c] Given by SEM

[d] Evidenced by XRD

[e] Synthesized in fluoride media at an F/Si ratio of 1

Ti-MWW-PI. They have different response to the mild acid treatment. The mild acid treatment over Ti-MWW-HMI resulted in 3D Ti-MWW structure other than Ti-MCM-56 structure (Table 3.2).To find the real reason for the failure of post-synthesis of Ti-MCM-56 form Ti-MWW-HMI, another Ti-MWW lamellar precursor was also prepared with PI as the SDA in the F⁻ medium, denoted as Ti-MWW-PI-F. The lamellar precursor of Ti-MWW-PI-F possesses the same organic ammonium molecules but has much larger particle sizes. Similar to Ti-MWW-HMI, only 3D Ti-MWW structure can be formed when Ti-MWW-PI-F was subjected to the mild acid treatment (Table 3.2), which indicates that particle size other than SDA molecules is the real factor that affects the construction of Ti-MCM-56 structure. In the case of Ti-MWW precursor with larger particle size, MWW sheets is more likely to keep the original stacking style even when part of the pillaring organic molecules are removed. This post-synthesis method can also be used to prepare other metallosilicates with MCM-56 structure.

With partial MWW layers exposed outside, MCM-56 showed an obvious uptake in the high-pressure region in the N_2 adsorption isotherms (Fig. 3.8), indicating larger external surface area than 3D Ti-MWW. In contrast, Ti-MCM-56 has lower adsorption amount than Ti-MCM-22 in the pressure region lower than $P/P_0 = 0.4$, which was caused by the interlayer blocking due to the deconstruction of Ti-MCM-56. A further calculation showed that Ti-MWW-56 have smaller micropore area but larger external surface area than 3D Ti-MWW. Therefore, it can be concluded that the increase of external surface area is at the expense of deconstructing the interlayer micropores. The enhanced external surface area can also be proved by the IR spectra. As shown in Fig. 3.9, the band of 3745 cm^{-1} assigned to terminal silanols was more intense for Ti-MCM-56 compared with 3D Ti-MWW, indicating larger part of MWW layers is exposed outside. Those bands around 3680 and 3500 cm^{-1} due to the defects of deboronation was much smaller for Ti-MCM-56, because only mild acid treatment was adopted in preparing Ti-MCM-56 zeolite.

Figure 3.10 shows the possible mechanism for the formation of partially del-aminated MCM-56 in comparison to 3D Ti-MWW. Transmission electron

Fig. 3.8 N_2 adsorption isotherms of 3D Ti-MWW (*a*) and Ti-MCM-56 analogue (*b*). Reprinted from Ref. [9], Copyright 2008, with permission from Elsevier

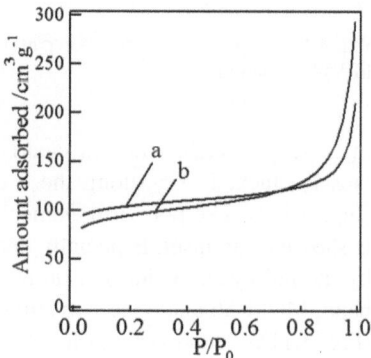

Fig. 3.9 IR spectra in region of hydroxyl stretching vibration of 3D Ti-MWW (*a*) and Ti-MCM-56 analogue (*b*). Reprinted from Ref. [9], Copyright 2008, with permission from Elsevier

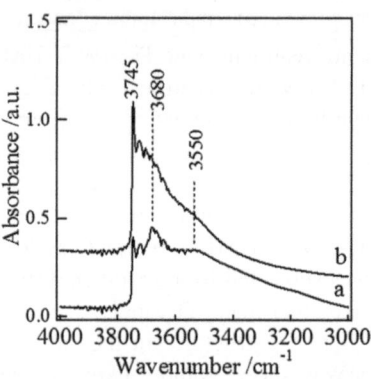

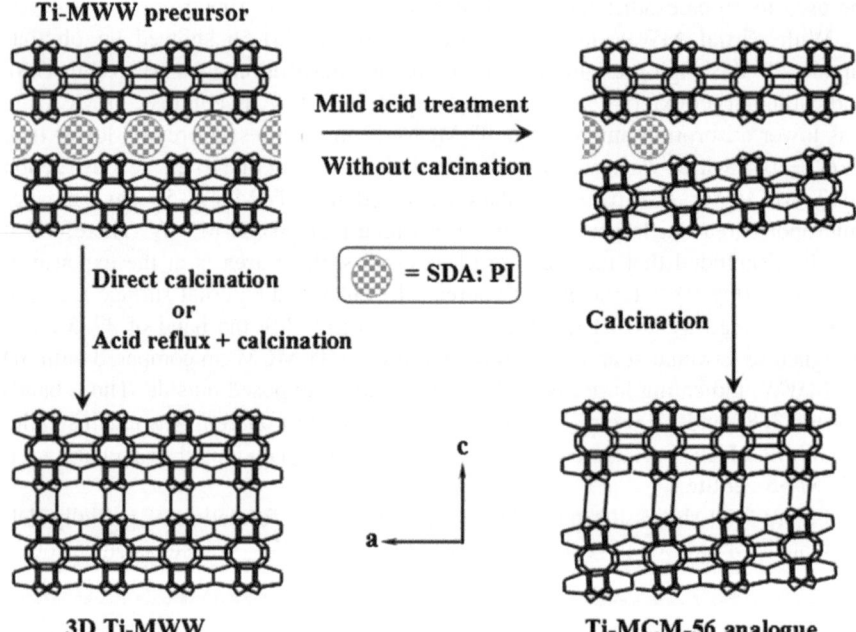

Fig. 3.10 Preparation process for partially delaminated Ti-MCM-56 in comparison to 3D-Ti-MWW formation

microscopy (TEM) images clearly show that the MWW layers (or sheets) are closely stacked very along the *c* direction (i.e., the direction of layer stacking) (Fig. 3.11a). The pore arrangement fitting in the MWW structure model perfectly is shown as an inset. Especially, the array of the intralayer 10-MR pores indicated by the red cycles is located on the same positions along the *c* axis. On the other hand, MCM-56 analogue was still composed of the collection of the MWW layers (Fig. 3.11b). However, the layer stacking mode lacks the regularity given by the

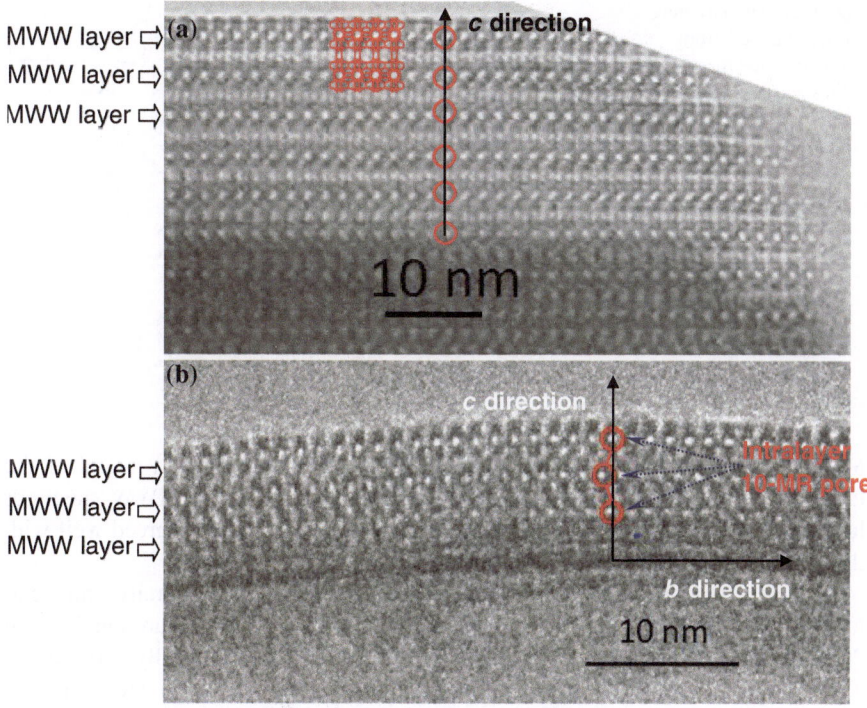

Fig. 3.11 HRTEM images of 3D MWW and typical post-synthesized MCM-56

Fig. 3.12 The powder XRD pattern of experimental spectra (*a*) and DIFFaX simulated one of MCM-56 (*b*). The simulated pattern was obtained by shifting the layer along *b* axis for 0.3 unit

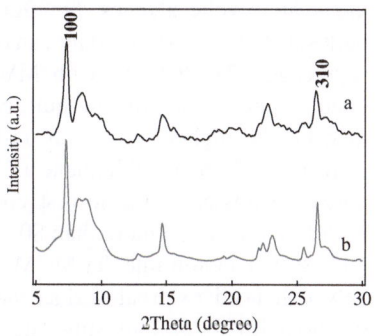

3D MWW structure. According to the positions of the intralayer 10-MR pores, the layers could be shifted along the *b* direction.

According to TEM observation, the simulation of the XRD pattern was performed using DIFFax. The post-synthesized MCM-56 analogue showed a pronounced peak broadening in the XRD pattern (Fig. 3.12a). It was found the crystals that are composed of two unit cells as well as shifted layers along *b* axis for 0.3 unit cell give a reasonable agreement with the simulated XRD pattern of

Fig. 3.13 The structure
model obtained from
simulation for the post-
sythesized MCM-56 with a
partially delaminated MWW
structure

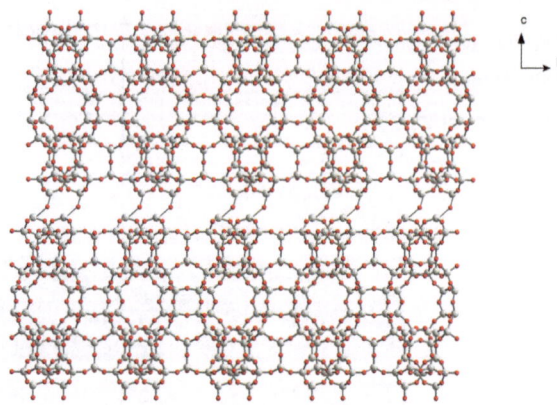

MCM-56 (Fig. 3.12b). On the basis of XRD pattern simulation, a model of the
layer arrangement was constructed as shown in Fig. 3.13. The MWW sheets or
layers are arranged in a shifted manner along *b* axis. The results agreed well with
the above-mentioned TEM image.

The as synthesized Ti-MWW precursor contain both tetrahedrally and octahedrally coordinated Ti species, identified by the adsorption bands around 220 and
260 nm in UV–Vis spectra. After the acid treatment, the octahedrally coordinated
Ti atoms are removed leaving only highly dispersed active Ti atoms in the
framework. The catalytic performance of Ti-MCM-56 was characterized in the
epoxidation of various alkenes with the oxidant of TBHP. Ti-MCM-56 showed
higher activity than TS-1, Ti-Beta, and 3D Ti-MWW in the epoxidation of both
linear and cyclic alkenes. For 3D Ti-MWW zeolite, epoxidation reactions with
bulk substrates or/and oxidant can only take place in the outside pockets and inside
supercages. Ti-MCM-56 with MWW sheets exposed outside can provide more
open reaction space for the bulk reactions, resulting in superior catalytic performance. The higher activity of Ti-MCM-56 was verified in the oxidative desulfurization of dibenzothiophene (DBT). Figure 3.14 shows that TS-1 with small
pore size possesses the lowest conversion rate. Ti-Beta, 3D Ti-MWW and Ti-
MCM-56 converted more than 80 % of the substrates with Ti-MCM-56 having the
highest conversion rate. Ti-MCM-56, with well-preserved MWW sheets and largely enhanced external surface area, is a promising catalyst for the oxidation
reactions involving bulk substrates as well as bulk oxidants.

3.5 Pillaring of Ti-MWW

Swelling of lamellar MWW precursor resulted in an expanded structure with
surfactant molecules intercalated in the interlayer space like a sandwich. Thus, the
interlayer space was greatly enhanced with a *d* value of 3.9 nm. The large

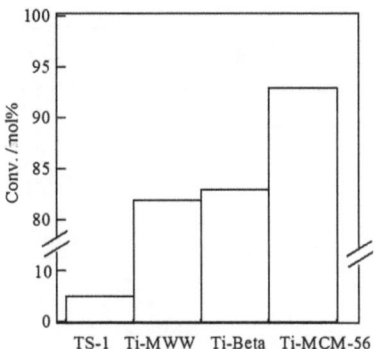

Fig. 3.14 A comparison of oxidative desulfurization of dibenzothiophene between various titanosilicates. Conditions: cat., 0.1 g; solvent (acetonitrile for Ti-MWW, Ti-MCM-56 and Ti-Beta, MeOH for TS-1), 10 mL; model light oil (DBT, 1000 μL mL^{-1} in isooctane), 10 mL; H_2O_2 (30 %),136 lL; n (H_2O_2):n (DBT) = 4; temp., 343 K; time, 3 h. Reprinted from Ref. [9], Copyright 2008, with permission from Elsevier

interlayer distance was lost when the swollen material was subjected to calcination. SiO_2 pillars can be used to preserve the open space between layers even upon calcination. Pillaring the swollen MWW structure results in a new derivate called MCM-36, which is made up of both micropores in its crystalline layers and mesopores in the interlayer space. This pillaring technology has achieved great success in aluminosilicate MWW precursor and lead to a new pillared aluminosilicate with higher specific surface area and good accessibility for relatively large molecules [30, 31]. This method was also applied to pillar lamellar titanosilicate MWW precursor to Ti-containing MCM-36 structure [10], which was expected to show higher activity in liquid phase oxidation reactions. Figure 3.15 shows the preparation procedures for microporous/mesoporous hybrid Ti-MCM-36.

The pre-swelling process was carried out by refluxing the mixture of Ti-MCM precursor, cetyltrimethylammonium chloride (CTMACl), and TPAOH. Then, TEOS was introduced into the mixture to provide silica source, followed by hydrolysis and calcination to yield siliceous pillars between the layers. It is suggested that a mesoporous region between the layers is created by polymeric silica pillars formed by TEOS hydrolysis.

After swelling and pillaring, the 002 peaks disappeared while a new diffraction peak with $2\Theta = 1–2°$ appeared indicating that interlayer space was expanded and mesopore region formed during this process. 3D Ti-MWW showed typically type I nitrogen adsorption isotherm characteristic of micropore channels, with a specific surface area of 471 cm^2 g^{-1}. As for Ti-MCM-36, the isotherm is characteristic of a hysteresis loop beginning at $P/P_0 = 0.4$ due to the presence of mesopores. The specific surface area of Ti-MCM-36 is 674 cm^2 g^{-1}, much higher than that of 3D-Ti-MWW.

As a catalyst, the state of Ti species in the framework of Ti-MCM-36 should be carefully identified before it is used to catalyze the oxidation reactions. The UV–Vis

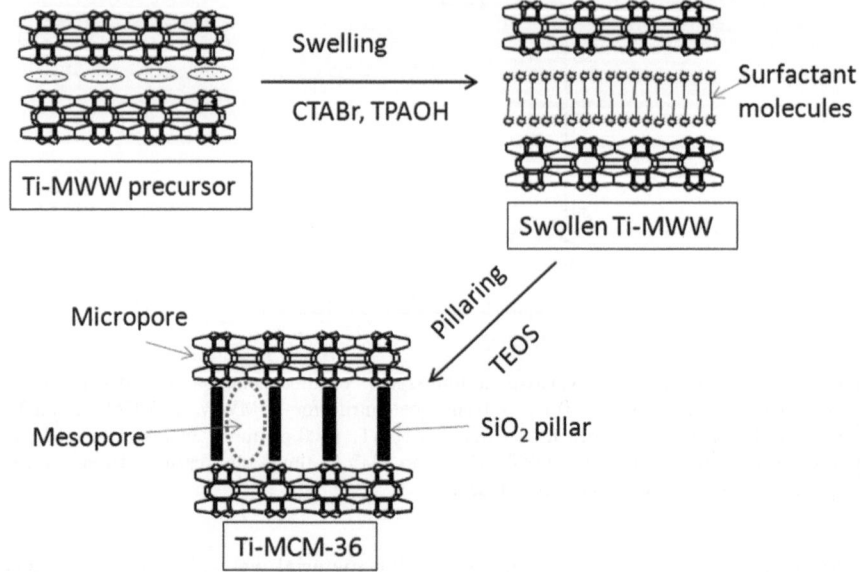

Fig. 3.15 Preparation procedure for Ti-MCM-36 microporous mesoporous hybrid

spectra of Ti-MCM-36 revealed that both active tetrahedral Ti atoms and inactive TiO$_2$ species are involved in this pillared material. UV–Vis spectra have also shown that the amount of anatase TiO$_2$ can be largely decreased when the Ti-MWW lamellar precursor was treated by a mild acid washing to remove the octahedrally coordinated Ti species in advance, because the octahedral Ti species can easily form anatase TiO$_2$ during the swelling and pillaring process.

For the epoxidation of 1-hexene, Ti-MCM-36 showed a higher TON value and H$_2$O$_2$ conversion than Ti-MWW [10]. The high activity of Ti-MCM-36 was believed to come from the contribution of mesopore region in the interlayer space. The interlayer 10-MR pores of 3D Ti-MWW were enlarged to mesopores favoring the accessibility of substrates to Ti active sites involved inside the framework.

3.5.1 Interlayer Silylation of Ti-MWW

Ti-MWW lamellar precursor with SDA molecules pillaring MWW sheets has a layer spacing of 26.93 Å. The removal of organic SDA molecules through calcinations caused condensation and dehydroxylation between the neighboring layers forming 10-MR pore channels in the interlayer space, which resulted in a decreased layer spacing of 25.09 Å. Once 3D MWW structure with interlayer 10-MR pore channels formed, the accessibility of bulky substrates to the supercages is seriously restricted. The above described post-treatment techniques including swelling, delamination, and pillaring are used to create more open reaction space. Then another interlayer expanded material called Ti-YUN-1 with interlayer newly formed 12-MR channels was obtained occasionally when the Ti-MWW lamellar precursor was subjected to an acid treatment [32, 33]. The above-mentioned Ti-MWW lamellar precursor was synthesized by the structural conversion method. This special phenomena only happened when the Si/Ti ratio was higher than 70. HRTEM images and computational approach revealed that the structure of Ti-YUN-1 was probably constructed through interlayer pillaring with monomeric Si or Ti species [34].

Inspired by the construction mechanism of Ti-YUN-1, silanes were introduced to silylate the Ti-MWW lamellar precursor in order to obtain an interlayer expanded structure [11, 12]. The silylation process was achieved by refluxing the mixture of silanes, Ti-MWW lamellar precursor and HNO_3 solution at 373 K for 20 h. Under such condition, silanes such as diethoxydimethylsilane (DEDMS) reacted with the silanol groups on the layer surface and then connected the up-and-down layers to form an interlayer expanded structure called IEZ-Ti-MWW. Unlike Ti-MWW lamellar precursor, the diffraction peaks related to the c axis remained intact when the silylated sample was subjected to calcination. Characteristic of well resolved 001 and 002 diffraction peaks, the XRD pattern of silylated sample was very similar with that of its corresponding lamellar precursor. However, well resolved diffraction peaks in the 2Θ region of 10–25° other than relatively broad ones, indicating ordered linkages between the MWW sheets may form during the silylation process instead of hydrogen bonding in the precursor. According to the XRD patterns, the layer spacing of IEZ-Ti-MWW was 27.58 Å, larger than both of the lamellar precursor and 3D Ti-MWW. However, the cell parameters along a and b axes for the three samples were almost the same, indicating silylation caused a structural change mainly along c axis.

Silylation involved removing some of the SDA molecules out of the interlayer space to make room for the silanes and then silanes diffusing into the interlayer space to react with the silanol groups on the layer surface. Therefore, the silylating conditions should be well controlled to match the above two dynamic processes well. Monomeric Si species come from the framework without additional Si source when constructing structure of Ti-YUN-1. In the contrary, silanes were added into the silylation procedure to provide enough Si source. Several silanes including TEOS, triethoxymethylsilane, trimethylethoxysilane and DEDMS were considered. To connect the up-and-down sheets of MWW by forming Si–O–Si linkages, the silylating agents should contain at least two active ethoxy groups. However, trimethylethoxysilane only has one active group and it was supposed to connect with one of the layers and then connected to the other by calcination. Table 3.3 shows the catalytic performance of IEZ-Ti-MWW prepared with different silanes. TEOS with four active groups cannot avoid self-condensation in such acidic aqueous solution forming amorphous species adhere to the surface, resulting in a decrease of crystallinity. As a result, IEZ-Ti-MWW showed the lowest activity in the epoxidation of cyclohexene. The other three silanes with at least one methyl group in the molecules can effectively prevent the self-condensation leading to

Table 3.3 The catalytic properties shown by 3D Ti-MWW and IEZ-Ti-MWW prepared with different silylating agents in the oxidation of cyclohexene[a]

Silylating agent	Cyclohexene oxidation				
	Conversion (mol%)	Product selectivity (mol%)		H_2O_2 (mol%)	
		Oxide	Others[b]	Conversion	Selectivity
None	4.1	32.5	67.5	5.6	69.6
TEOS	8.8	43.3	56.7	9.4	92.9
Triethoxymethylsilane	19.2	53.5	46.5	20.4	90.3
Trimethyethoxylsilane	19.8	54.6	45.4	21.0	94.3
DEDMS	20.7	57.1	42.9	19.9	98.3

Reprinted from Ref. [12], Copyright 2009, with permission from Royal Society of Chemistry
[a] Reaction conditions: cat., 0.05 g; cyclohexene, 10 mmol; H_2O_2, 10 mmol; MeCN, 10 mL; temp., 333 K; time, 2 h
[b] 2-Cyclohexene-1-ol, 2-cyclohexene-1-one, glycols

IEZ-Ti-MWW with much higher activity than the one prepared with TEOS, among which the one silylated by DEDMS showed the highest activity and selectivity.

In theory, every unit cell of MWW lamellar precursor has two pillaring sites, which means the amount of silanes should also be well controlled. It is supposed that too small amount of silanes would result in a disordered structure due to incomplete silylation while excess silane molecules would react with each other to form amorphous impurity attached on the layer surface. XRD patterns have showed that too small amount of silylation agents lead to IEZ-Ti-MWW with very weak 001 and 002 diffraction peaks. With the increase amount of the silylation agents used, the peaks become more intense, meaning a more ordered structure was obtained. IR spectra (Fig. 3.16) were also employed to monitor the silylation process. All the samples showed two bands at 2580 and 2926 cm^{-1}, assigned to the stretching vibration of CH_2 groups originating from the HMI molecules. The remaining HMI molecules were supposed to be involved in the intralayer 10-MR pores that were hard to be removed by mild acid washing in the silylation procedure. After silylation, two new bands appeared in the IR spectra at 2970 and 850 cm^{-1}, attributed by the CH_3 asymmetric stretching vibration and the rocking of CH_3 groups attached to Si, respectively. This can be taken as the evidence for the incorporation of silane molecules into the framework. The two bands became more intense when increasing the amount of silanes but leveled off when the weight ratio of silane to precursor is set at 0.15, which was in agreement with the theoretical optimal amount. The number of silane molecule that incorporated into every unit cell can be calculated based on the incorporated carbon and the chemical composition of the sample. According to the chemical analysis, the number of inserted DEDMS molecules increased from 1.9 to 3.1 molecules per unit cell when weight ratio of silane to precursor increased from 0.05 to 0.30. It should be noted that the number of incorporated DEDMS almost unchanged when the weight ratio of added DEDMS to precursor was larger than 0.15, which is in

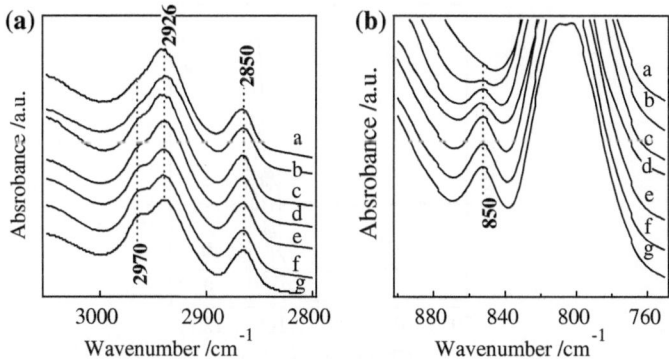

Fig. 3.16 FT-IR spectra of (*a*) acid treated Ti-MWW with 2 M HNO₃, and IEZ-Ti-MWW prepared by silylation with (*b*) 0.025 g, (*c*) 0.05 g, (*d*)0.07 g, (*e*) 0.1 g, (*f*) 0.15 g, (*g*) 0.3 g of DEDMS per gram of the precursor in 2 M HNO₃ without calcination. The silylation resulted in bands at 2970 and 850 cm⁻¹ assigned to the asymmetric vibration CH₃ groups and the rocking vibration of Si–CH₃, respectively. Reprinted from Ref. [12], Copyright 2009, with permission from Royal Society of Chemistry

agreement with the results of IR spectra and XRD patterns. These samples that prepared with different amount of silane molecules also behaved differently in epoxidation of 1-hexene and cyclohexene. The linear type alkene of 1-hexene is small enough to diffuse through the 10-MR pore entrance to reach the intracrystal supercages without constrains. Therefore, there was no obvious difference of activity between the 3D Ti-MWW and the silylated samples. However, the epoxidation of the bulky cyclohexene molecules can only take place at Ti active sites in the side pockets and supercages. As for the interlayer connected 3D Ti-MWW structure, cyclohexene cannot access the Ti species in the supercages and only can be catalyzed by the Ti species in the outer-surface side pockets. With the expanded structure, IEZ-Ti-MWW provided more opener reaction space for the bulky cyclohexene molecules and therefore increased the activity. Among all the silylated samples, the one prepared with the optimal amount of DEDMS molecules showed the highest activity with the TON value of 233 in catalyzing the epoxidation of cyclohexene.

Monomeric Si species with two methyl groups were inserted into the framework after silylation, which has been verified by IR spectra. ^{29}Si NMR spectra (Fig. 3.17) also proved the introduction of methyl groups with the appearance of resonance band at -12 ppm assigned to $Si(CH_3)_2(OSi)_2$groups (D^2). D^2 groups were also detected by ^{13}C NMR spectra with the resonance band at -2.0 ppm. Along with the appearance of D^2 band, the band at -103 ppm attributed to silanol groups suffered a great decrease, indicating DEDMS molecules consumed the silanos on the up-and-down surface and formed D^2 groups as pillars in the interlayer space. The obtained organic–inorganic hybrid material was then calcined to change the methyl groups to hydroxyl groups, yielding Q^2 groups in the structure with a resonance band at -90 ppm in the ^{29}Si NMR spectrum

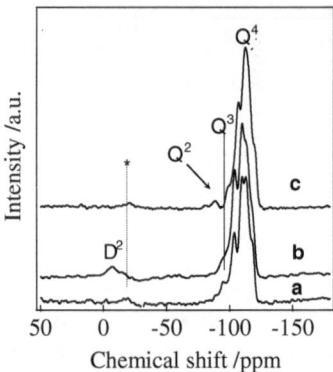

Fig. 3.17 ^{29}Si MAS NMR spectra of (a) Ti-MWW lamellar precursor, (b)after silylation with DEDMS in 2 M HNO$_3$ without calcination, and (c)after further calcination at 823 K. The silylation consumed a portion of the silanol groups (Q^3) and developed at -12 ppm a signal assigned to =Si–(CH$_3$)$_2$ groups (D^2) instead, which were converted to =Si(OH)$_2$ groups (Q^2) around -90 ppm after calcination. Reprinted from Ref. [12], Copyright 2009, with permission from Royal Society of Chemistry

(Fig. 3.17c). The appearance of D^2 groups after silylation and Q^2 groups upon calcination again identified the fact that silanes were successfully inserted into the MWW structure.

TEM images are supposed to be the most direct technique to elucidate the real structure of IEZ-Ti-MWW zeolite. As shown in Fig. 3.18a, the TEM images taken along 001 incidence showed ordered arrays of 12-MR side cups were the same as 3D Ti-MWW zeolite, indicating silylation did not alter the structure within the MWW planes. The other TEM image (Fig. 3.18b) collected along 100 incidence showed an ordered pore structure along the stacking direction, suggesting that IEZ-Ti-MWW also had 3D connected structure analogue to 3D Ti-MWW. The layer spacing of 27.4 Å calculated from the TEM images of IEZ-Ti-MWW indicated that the structure was expanded compared to the conventional 3D Ti-MWW structure, which was the result of ordered silylation process. Although the real structure of IEZ-Ti-MWW should be identified by synchrotron X-ray diffraction, the new 12-MR pores constructed by two inserted Si atoms from silanes and the other ten Si atoms from the up-and-down layers were formed during silylation and the following calcination.

IEZ-Ti-MWW with interlayer expanded structure is able to provide more wide diffusion path for bulky molecules to access the intracrystal supercages. And it is expected to have better performance than the conventional 3D Ti-MWW catalyst. IEZ-Ti-MWW and 3D Ti-MWW with different Ti contents were both used to catalyze the epoxidation of cyclohexene. With the increase of Ti content, the conversion of cyclohexene showed a great enhance over IEZ-Ti-MWW while only limited increase was observed as for 3D-Ti-MWW zeolite. With diffusion constrains, the Ti active sites in 3D Ti-MWW cannot be fully used by cyclohexene molecules. Therefore, IEZ-Ti-MWW showed superior performance than the 3D

Fig. 3.18 TEM images of
IEZ-Ti-MWW taken along
(**a**) [001] incidence and
(**b**) [100] incidence. FDs are
also inset in the images

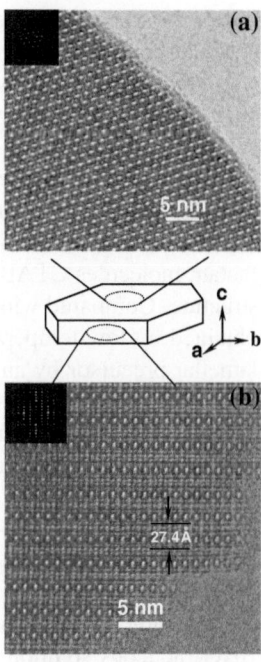

Ti-MWW with comparable Ti content. It has also been proved that lamellar MWW precursors with other metal atoms in the framework can also be silylated to interlayer expanded structure with higher performance in either reactions or adsorptions.

3.5.2 Synthesis of Ti Containing Interlayer Expanded MWW Structure Via a Combination of Silylation and Ti Insertion

Silylation of Ti-MWW lamellar precursor with monomeric silanes such as DEDMS constructed an interlayer expanded structure with 12-MR channels in the interlayer space. Although the pore size of the newly formed 12-MR pores is larger than that of the 10-MR pores in the conventional 3D Ti-MWW zeolite, the interlayer structure still showed diffusion constrains when very bulky substrates were involved. To pursue larger interlayer space, dipodal silanes were introduced to construct the interlayer structure and were expected to result in 14-MR pores. Unfortunately, the silylation with dipodal silanes led to an interlayer expanded structure with the same layer spacing as that obtained by monomeric silanes, implying an unsuccessful silylation process. It was supposed that the interlayer space of the lamellar precursor was too narrow to allow the dimeric silanes to diffuse in. Then the pre-swelling process was considered to increase the interlayer space large enough to accommodate the dimeric silane molecules. Unlike the one-step silylation process with monomeric silanes, silylation process with dimeric silanes involved both swelling and silylation, during which Ti active sites were supposed to be severely affected. To obtain an active titanosilicate MWW with interlayer 14-MR pore channels, the insertion of titanium atoms was carried out after the interlayer expanded structure was constructed.

Borosilicate MWW was at first swollen in the mixture of TPAOH and surfactant molecules CTAB at room temperature to avoid dissolving the MWW layer structure. Compared with high-temperature swelling process, the swollen structure obtained by low-temperature process was reversible and can be changed back to lamellar precursor by an acid treatment, indicating this swelling process was very mild. TPAOH molecules were used to provide the alkaline conditions and cleave the interlayer connection of hydrogen bonding. To protect the MWW layers from desilication, the amount of TPAOH should be well controlled. The interlayer spacing showed a peak value of 4.5 nm when the amount of TPAOH was 2.5 g per gram of the precursor B-MWW zeolite. An overlap between the diffraction peaks of 101 and 102 appearing on the XRD patterns of swollen B-MWW zeolite can be taken as the clue for the success of this low-temperature swelling process. Then, the mixture of swollen B-MWW zeolite, 1,2-dichlorotetramethyldisilane, and HNO_3 aqueous solution was refluxed to yield interlayer expanded B-MWW zeolite, denoted as IEZ-MWW(SiSi). The mechanism of silylation with dimeric silanes was the same as that carried out with monomeric silanes. Therefore, the amount of dimeric silane should also be well controlled to obtain an ordered interlayer expanded structure without amorphous impurity. The layer spacing changed from 2.8 to 3.5 nm when the amount of silanes varied from 1 to 4 mmol per gram of swollen B-MWW zeolite. The layer spacing of the IEZ-Ti-MWW zeolite obtained with monomeric silane was 2.75 nm and the length of Si–Si bond was 2.332 Å, indicating the most possible layer spacing of IEZ-MWW(SiSi) zeolite should be about 3.0 nm. IEZ-MWW(SiSi) zeolite obtained with the 2 mmol silanes per swollen B-MWW zeolite possessed the most close layer spacing with the theory value and the amount was taken as the optimal dosage (Fig. 3.19). The extreme large layer spacing was supposed to be caused by the condensation of the silane molecules between the interlayer spacing when excess silane molecules were added to the silylation mixture, which was the same case as the TEOS molecules in pillaring process. The resonance bands belonging to the SDA molecules of HMI can be observed by ^{13}C NMR spectra. After low-temperature swelling process, the resonance bands assigned to HMI became weaker because part of them were removed and some new resonance bands assigned to CTAB molecules appeared, indicating the intercalation of surfactant molecules. The resonance bands assigned to HMI almost disappeared and the bands belonging to CTAB molecules were weaker after the swollen B-MWW zeolite was subjected to silylation process. Thus, the acid washing during silylation process can only remove part of the surfactant molecules. Besides, a new band at round 0 ppm appeared after silylation, which can be attributed to the methyl

Fig. 3.19 XRD patterns of
as-synthesized B-MWW (*a*),
after swelling with CTABr at
room temperature (*b*), and
after silylation with ClMe₂Si-
SiMe₂Cl silane,
IEZ-MWW(SiSi) (*c*)

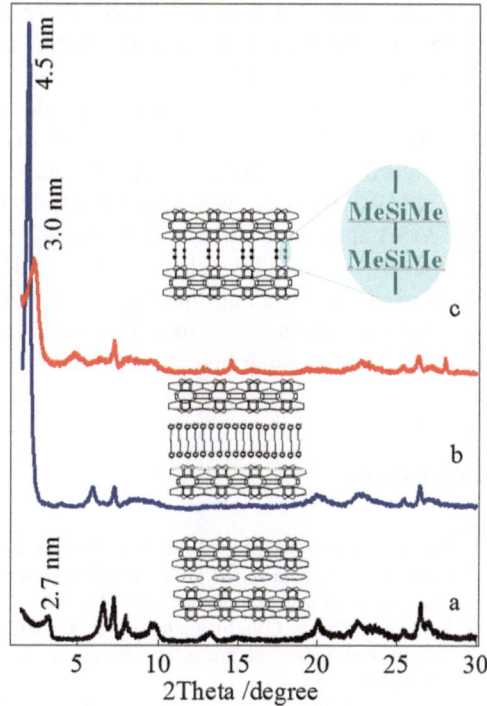

Fig. 3.20 The results of
epoxidation of cyclohexene
by different MWW materials

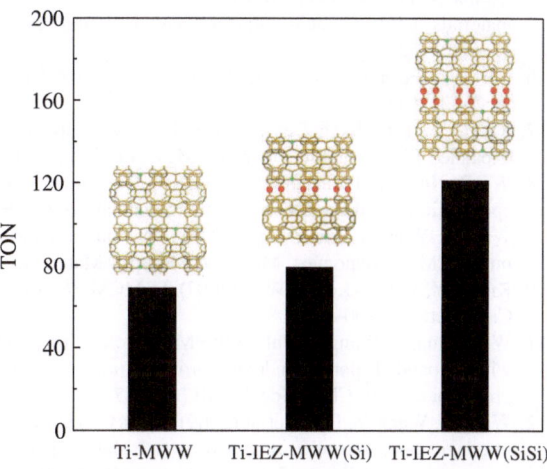

groups in the silane molecules connected to the up-and-down layers. Then, the
following calcination at 723 K removed almost all the surfactant molecules in the
IEZ-MWW(SiSi) structure while a new band around −6 ppm appeared, indicating
a new C species formed upon calcination. Some of the methyl groups in silane

molecules intercalated in the interlayer space would be calcined to hydrogen groups, leaving $(OH)(CH_3)Si(OSi)_2$ species in the structure corresponding to the resonance band at round -6 ppm.

After the well-ordered interlayer expanded structure with newly formed 14-MR pore channels has been constructed, the Ti insertion was realized by a liquid–solid phase reaction with H_2TiF_6 as the Ti source. With more opener reaction space, Ti-IEZ-MWW(SiSi) showed a higher activity than both 3D Ti-MWW and IEZ-Ti-MWW(Si) analogues with smaller interlayer pore channels especially when the reactions involved bulky substrates like cyclohexene (Fig. 3.20). Thus, the combination of pre-swelling, interlayer silylation with dipodal silane and post-isomorphous substitution of transition metal cations would lead to promising redox catalysts for processing bulky molecules.

References

1. Zeolite structure database (2013) http://www.iza-online.org
2. Choi M, Na K, Kim J et al (2009) Stable sigle-unit-cell nanosheets of zeolite MFI as active and long-lived catalysts. Nature 461:246–250
3. Roth WJ, Shvets OV, Shamzhy M et al (2011) Postsynthesis transformation of three-dimensional framework into a lamellar zeolite with modifiable architecture. J Am Chem Soc 133:6130–6133
4. Verheyen E, Joos L, Van Havenbergh K et al (2012) Design of zeolite by inverse sigma transformation. Nat Mater 11:1059–1064
5. Corma A, Diaz U, Domine ME (2000) New aluminosilicate and titanosilicate delaminated materials active for acid catalysis, and oxidation reactions using H_2O_2. J Am Chem Soc 122:2804–2809
6. Corma A, Fornés V, Díaz U (2001) ITQ-18 a new delaminated stable zeolite. Chem Commun 24:2642–2643
7. Xu H, Yang B, Jiang J et al (2013) Post-synthesis and adsorption properties of interlayer-expanded PLS-4 zeolite. Micropor Mesopor Mater 169:88–96
8. Wu P, Nuntasri D, Ruan J et al (2004) Delamination of Ti-MWW and high efficiency in epoxidation of alkenes with various molecular sizes. J Phys Chem B 108:19126–19131
9. Wang L, Wang Y, Liu Y et al (2008) Post-transformation of MWW-type lamellar precursors into MCM-56 analogues. Micropro Mesopro Mater 113:435–444
10. Kim S-Y, Ban H-J, Ahn W-S (2007) Ti-MCM-36: a new mesoporous epoxidation catalyst. Catal Lett 113:160–164
11. Wu P, Ruan J, Wang L et al (2008) Methodology for synthesizing crystalline metallosilicates with expanded pore windows through molecular alkoxysilylation of zeolitic lamellar precursors. J Am Chem Soc 130:8178–8187
12. Wang L, Wang Y, Liu Y et al (2009) Alkoxysilylation of Ti-MWW lamellar precursors into interlayer pore-expanded titanosilicates. J Mater Chem 19:8594–8602
13. Schreyeck L, Caullet P, Mougenel JC et al (1996) PREFER: a new layered (alumino) silicate precursor of FER-type zeolite. Micropor Mater 6:259–271
14. Ikeda T, Akiyama Y, Oumi Y et al (2004) The topotactic conversion of a novel layered silicate into a new framework zeolite. Angew Chem Int Ed 43:4892–4896
15. Wang YX, Gies H, Marler B et al (2005) Synthesis and crystal structure of zeolite RUB-41 obtained as calcination product of a layered precursor: a systematic approach to a new synthesis route. Chem Mater 17:43–49

16. Wang L, Liu Y, Xie W et al (2008) Improving the hydrophobicity and oxidation activity of Ti-MWW by reversible structural rearrangement. J Phys Chem C 112:6132–6138
17. Tatsumi T, Koyano KA, Igarashi N (1998) Remarkable activity enhancement by trimethylsilylation in oxidation of alkenes and alkanes with H_2O_2 catalyzed by titanium-containing mesoporous molecular sieves. Chem Commun 3:325–326
18. Wu P, Tatsumi T, Komatsu T et al (2002) Postsynthesis, characterization, and catalytic properties in alkene epoxidation of hydrothermally stable mesoporous Ti-SBA-15. Chem Mater 14:1657–1664
19. Yamamoto K, Sakata Y, Nohara Y et al (2003) Organic–inorganic hybrid zeolites containing organic frameworks. Science 300:470–472
20. Asefa T, MacLachan MJ, Coombs N et al (1999) Periodic mesoporous organosilicas with organic groups inside the channel walls. Nature 402:867–871
21. Wu P, Komatsu T, Yashima T et al (1995) IR and MAS NMR studies on the incorporation of aluminum atoms into defect sites of dealuminated mordenites. J Phys Chem 99:10923–10931
22. Nagy JB, Gabelica Z (1982) A cross-polarization magic-angle-spinning ^{29}Si-n.m.r. indentification of the sianol group resonance in ZSM-5 zeolites. Chem Lett 11:1105–1108
23. Boxhoorn G, Kortbeek AGTG, Hays GR et al (1984) A high-resolution solid-state 29Si n.m.r. study of ZSM-5 type zeolites. Zeolites 4:15–21
24. Yamamura M, Chaki K, Wakatsuki T et al (1994) Synthesis of ZSM-5 zeolite with small crystal size and its catalytic performance foe ethylene oligomerization. Zeolite 14:643–649
25. Roth WJ, Vartuli JC (2002) In: Sayari A, Jaroniec M (eds) Nanoporous materials III. Stud Surf Sci Catal, Vol 141. Elsevier, New York, p 273
26. Corma A, Díaz U, Fornés V et al (1999) Ti/ITQ-2, a new material highly active and selective for the epoxidation of olefins with organic hydroperoxides. Chem Commun 9:779–780
27. Wang Y, Liu Y, Wang L et al (2009) Postsynthesis, characterization, and catalytic properties of aluminosilicates analogous to MCM-56. J Phys Chem C 113:18753–18760
28. Juttu GG, Lobo RF (2000) Characterization and catalytic properties of MCM-56 and MCM-22 zeolites. Micropor Mesopor Mater 40:9–23
29. Fung AS, Lawton SL, Roth WJ (1994) US Patent 5362697
30. He YJ, Nivarthy GS, Eder F et al (1998) Synthesis, characterization and catalytic activity of the pillared molecular sieve MCM-36. Micropor Mesopor Mater 25:207–224
31. Maheshwari S, Martínez C, Portilla MT et al (2010) Influence of layer structure preservation on the catalytic properties of the pillared zeolite MCM-36. J Catal 272:298–308
32. Fan W, Wu P, Namba S et al (2004) A titanosilicate that is structurally analogous to an MWW-type lamellar precursor. Angew Chem Int Ed 43:236–240
33. Fan W, Wu P, Namba S et al (2006) Synthesis and catalytic properties of a new titanosilicate molecular sieve with the structure analogous to MWW-type lamellar precursor. J Catal 243:183–191
34. Ruan J, Wu P, Slater B et al (2005) Structure elucidation of the highly active titanosilicate catalyst Ti-YNU-1. Angew Chem 117:6877–6881

Chapter 4
Catalytic Properties of Ti-MWW in Selective Oxidation Reactions

Abstract Ti-MWW/H$_2$O$_2$ has formed a new catalytic system that is useful to a variety of liquid-phase oxidation reactions such as the epoxidation of linear alkenes, cyclic alkenes, functionalized alkenes, the ammoximation of ketones, the oxidation of amines and sulfides, etc. Using H$_2$/O$_2$ instead of H$_2$O$_2$ aqueous solution as oxidant, Au/Ti-MWW is applied in the gas-phase epoxidation of propylene to propylene oxide. In this chapter, the catalytic properties of Ti-MWW in these reactions will be discussed in detail in comparison to other titanosilicates.

Keywords Catalysis · Ti-MWW · Titanosilicate · Epoxidation · Ammoximation · Selective oxidation

4.1 Overview

The discovery of TS-1 zeolite with the MFI structure by Taramasso et al., in 1983 [1], is a landmark in the heterogeneous catalysis of selective oxidation. TS-1 is the first molecular sieve containing a redox metal cation, Ti^{4+}, in the framework and showing excellent activity and selectivity for many industrial important organic selective oxidation reactions with H$_2$O$_2$ aqueous solution. Many research evidences have confirmed that tetrahedrally coordinated, isolated Ti^{4+} ions in the MFI structure are the origin of catalytic behavior [2, 3]. Typical reactions, catalyzed by TS-1, include epoxidation of alkene, hydroxylation of alphatic or aromatic compounds, ketone ammoximation, and conversion of alkanes to alcohols and ketones [4]. So far, there are three kinds of commercial plants using TS-1 as catalyst known in operation, they are hydroxylation of phenol to dihydroxy benzenes [5], ammoximation of cyclohexanone [6], and selective oxidation of propylene to propylene oxide (PO) [7]. These successful applications have been universally considered as a new era of green chemical industrial processes based on zeolite catalysis. However, there are also many drawbacks within industrial applications. The medium pore of TS-1 is too small, which restricts its application to both substrates and oxidant with relatively small molecular size. Besides, the high cost

P. Wu et al., *MWW-Type Titanosilicate*,
SpringerBriefs in Green Chemistry for Sustainability,
DOI: 10.1007/978-3-642-39115-6_4, © The Author(s) 2013

of manufacturing of TS-1 catalyst is another major constraint. To overcome these drawbacks, many titanosilicates such as TS-2 [8], Ti-Beta [9, 10], Ti-ZSM-12 [11], Ti-MOR [12], Ti-ITQ-7 [13], Ti-MWW [14], and Ti-containing mesoporous materials such as Ti-MCM-41 [15] and Ti-SBA-15 [16] have been prepared, and also their catalytic properties have been investigated.

Among these catalysts, Ti-MWW is expected to be served as a candidate that is highly active and stable. Besides 2D sinusoidal channels of 10-MR running throughout the structure parallel to the *ab*-plane, the MWW structure contains an independent channel system, which is comprised of 12-MR supercages [17, 18]. The supercages turn to be pockets or cup moieties at the crystal exterior. The unique pore system of intracrystalline supercages and abundant pockets covering the hexagonal faces of thin crystals have been proved to serve as an open reaction space in many reactions. As one of the most representative lamellar zeolites, Ti-MWW could be converted into fully delaminated material Del-Ti-MWW [19], partially delaminated materials Ti-MCM-56 [20], interlayer-pillared material Ti-MCM-36 [21], and interlayer-expanded material Ti-YNU-1 [22, 23] and IEZ-Ti-MWW [24, 25]. In comparison with Ti-MWW, these titanosilicates generally preserve the basic structure of MWW zeolite but also possess more open pores or more accessible active sites to the reactants.

Till date, Ti-MWW shows a unique solvent effect and high catalytic activities for the epoxidation of linear alkenes and various functional alkenes with H_2O_2 aqueous solution [26–33], and it is capable of catalyzing the ammoximation of ketones and aldehydes such as methyl ethyl ketone and cyclohexanone, with a ketone conversion and oxime selectivity both as high as 99 % [34, 35]. Moreover, Ti-MWW is extremely robust and shows a longer lifetime than the conventional TS-1 catalyst in the liquid-phase ammoximation of cyclohexanone with ammonia and hydrogen peroxide in a continuous slurry reactor [36]. Due to the deactivation in the ammoximation of cyclohexanone in the continuous slurry reactor, the core-shell-structured Ti-MWW@meso-SiO$_2$ have been fabricated and exhibits significantly prolonged lifetime in comparison to the parent Ti-MWW catalyst, which would develop a clean process for producing cyclohexanone oxime [37]. Ti-MWW is also an efficient catalyst in the oxidation of amines [38] and sulfides [39]. Using H_2/O_2 instead of H_2O_2 aqueous solution as oxidant, Au/Ti-MWW, which is achieved by deposition–preposition method, is applied in the gas-phase epoxidation of propylene to PO [40].

Ti-MWW/H_2O_2 has formed a new catalytic system that is useful to a variety of liquid-phase oxidation reactions. In this section, the catalytic properties of Ti-MWW in these reactions will be discussed in detail in comparison to other titanosilicates.

4.2 Epoxidation of Alkenes Over Ti-MWW

The success of TS-1 with MFI structure, in catalyzing the selective epoxidation of various alkenes in the liquid-phase using hydrogen peroxide as an oxidant, is a great milestone in the field of heterogeneous catalysis [41, 42]. Ti-Beta [9, 10],

Ti-ITQ-7 [13], Ti-MCM-41 [15], and Ti-MCM-48 [43] were developed to over-come the diffusion constraint in the application of TS-1. However, these catalysts are still less active than TS-1 in the reaction of reactants with small molecular diameter. On the contrary, Ti-MWW not only showed high activity in the selective epoxidation of propylene, allyl chloride (ALC), allyl alcohol (AAL), and diallyl ether, but also exhibited excellent activity in the epoxidation of cycloalkenes [30]. Besides, Ti-MWW showed a unique solvent effect on catalytic performance, which is different from that of TS-1 and Ti-Beta.

4.2.1 Epoxidation of Propylene Over Ti-MWW

PO is an important bulk chemical which is widely used as the intermediate in the production of polyurethane polyols, glycol ethers, dipropylene glycol, industrial polyglycols, lubricants, surfactants, oil demulsifiers, and isopropanolamines, and is also used as a solvent and soil sterilant [44, 45]. Conventional manufacturing processes for PO are mainly chlorohydrin process, TBHP process (Halcon method), and ethylbenzene hydroperoxide process (Shell method). The by-pro-duction of stoichiometric amount of waste salts as well as a large quantity of chlorine-containing water in the chlorohydrin process cause serious problems such as equipment corrosion and environmental pollution. In the Halcon process, the desired PO is co-produced together with *tert*-butanol, which appears to be the main limitation or disadvantage of this method. The process of ethylbenzene hydro-peroxide also cannot avoid the production of by-product like ethylbenzene, and it suffers the disadvantage of higher energy consumption. In a word, all these PO processes co-produce a large amount of wastewater together with organic and/or inorganic by-products, which need to be recycled or disposed [46]. The direct epoxidation of propylene with hydrogen peroxide to PO is thus a more attractive route owing to the easy handling and cleanness of the process. BASF thus has developed a TS-1/H_2O_2/methanol catalytic system for PO production, and has announced its commercialization in 2009 [7, 47]. Nevertheless, this innovative process may still be facing some drawbacks such as the zeotropic problem of methanol and PO in routine separation, and moreover, the formation of solvolysis by-products of glycol ethers as a result of ring-open reactions of PO with methanol.

The liquid-phase epoxidation of propylene with H_2O_2 aqueous solution to produce PO have been performed over Ti-MWW and other titanosilicates with different topology [33]. The catalytic results are listed in Table 4.1. The epoxidation of propylene gives PO as a main product. PO undergoes further reactions such as solvolysis and successive oxidation. As shown in Scheme 4.1, the solvolysis of PO by alcohol or water over acid sites of titanosilicates produces glycol ethers and diol, while the deep oxidation probably gives alcoholic hydroperoxides. Nevertheless, the by-products detected were mainly diol and monomethyl ethers (MME). The reac-tions were carried out in aprotic acetonitrile for Ti-MWW, Ti-Beta, and Ti-MOR,

Table 4.1 The results of propylene epoxidation over various titanosilicates in different solvents[a]. Reprinted from Ref. [33], Copyright 2007, with permission from Elsevier

No.	Catalyst[b]	Solvent	Amount (g)	n_{PO} (mmol)	PO selectivity[c] (%)	X_{H2O2} (%)	U_{H2O2} (%)
1	Ti-MWW(38)	CH_3CN	0.10	19.9	99.9	75.8	88.0
2	Ti-MWW(38)	CH_3CN	0.15	23.9	99.9	83.0	94.0
3	TS-1(42)	CH_3CN	0.10	7.2	99.8	25.0	96.0
4	TS-1(42)	CH_3OH	0.15	19.7	97.8	76.4	86.0
5	Ti-Beta(76)	CH_3CN	0.15	0.9	90.0	5.8	49.0
6	Ti-MOR(90)	CH_3CN	0.15	1.1	88.0	4.0	87.0

[a] Conditions: H_2O_2, 30 mmol; solvent, 10 g; temp., 313 K; pressure, 0.25 MPa; time, 1 h
[b] The number in parentheses indicates the Si/Ti molar ratio
[c] Propylene glycol (PG) and its monomethyl ethers (MME) were the main by-products

while in protic solvent of methanol for TS-1. The most suitable solvent for Ti-MWW was acetonitrile and for TS-1 was methanol. It is widely accepted that the use of different solvents results in different active species. Species I, with a stable 5-MR structure formed by the coordination of ROH (or H_2O) to Ti centers and hydrogen bonding to Ti-peroxo complex, is believed to be the active intermediates in protic alcohols, whereas species II is assumed to contribute to the oxidation of substrates in aprotic solvents (Scheme 4.2). Under the optimum reaction conditions, Ti-MWW showed a PO yield of 79.7 % and PO selectivity of 99.9 %. Although Ti-MWW contained a relatively high content of boron, the originated acid sites did not show great influence on the ring opening of PO in acetonitrile. Under the same reaction conditions of Ti-MWW except for solvent, TS-1 showed slightly lower PO yield,

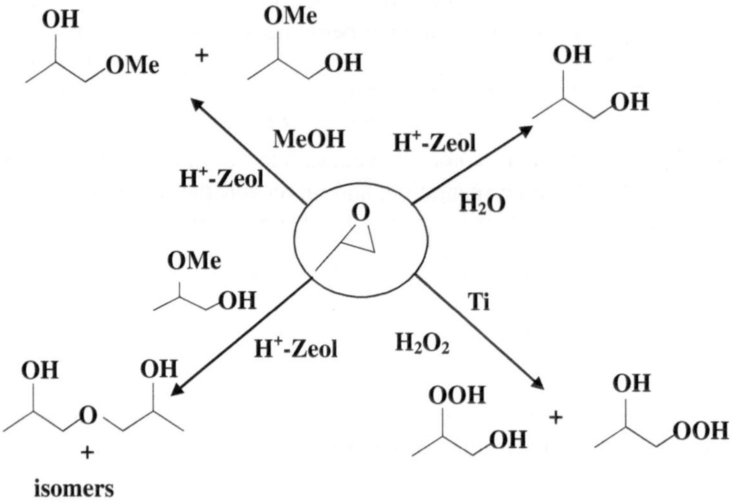

Scheme 4.1 Possible by-products produced in propylene epoxidation with H_2O_2

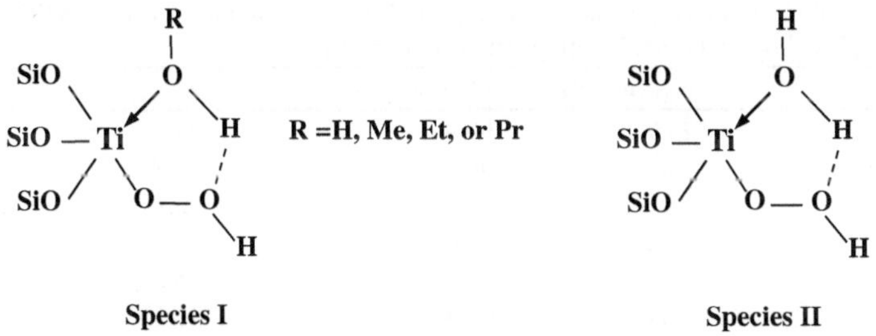

Species I   **Species II**

Scheme 4.2 Five-membered ring Ti hydroperoxo intermediate proposed in the liquid-phase oxidation

65.7 %, and PO selectivity of 97.8 %, which would be due to the unique pore system of MWW structure favoring the adsorption and access of substrate molecules to the Ti active sites. Ti-Beta and Ti-MOR exhibited much less active and selective than Ti-MWW and TS-1. The lower activity of Ti-Beta was found to be caused by a high density of connectivity defects, resulting in extreme hydrophilic properties. Although Ti-MOR had 12-MR pore, the large crystals as well as one-dimensional channels hinder the diffusion of reactants to active sites.

To investigate the solvent effect on the propylene epoxidation over Ti-MWW, the solvent of CH_3CN, acetone, CH_3OH, 1,2-dichloroethane, DMF, THF, 1,4-dioxane, and water have been used in this reaction [33]. The catalytic results are listed in Table 4.2. As reported previously, the most suitable solvent for the epoxidation of alkene over TS-1 is protic solvent of methanol. However, methanol and product PO are difficult to separate because they have similar boiling point and can easily form zeotropic compounds. Moreover, the protic solvent of methanol easily leads to the ring-opening reactions of PO to produce glycol ethers, which would decrease the selectivity of PO. However, the most suitable solvent for the epoxidation of propylene over Ti-MWW is aprotic solvent of CH_3CN. The PO yield of Ti-MWW decreased in the order of $CH_3CN >$ acetone $\approx$ 1,2-dichloroethane $> CH_3OH > H_2O >$ THF $>$ DMF.

The stability and reusability of Ti-MWW in the liquid-phase propylene epoxidation was compared with that of TS-1 (Fig. 4.1) [33]. The catalytic reactions were carried out in CH_3CN for Ti-MWW and in methanol for TS-1, respectively. The used catalyst was regenerated by washing with acetone and drying at 393 K and then used repeatedly in the next epoxidation reactions. After fourth runs, the activity of propylene epoxidation decreased to 0.08 for Ti-MWW and 0.06 mol-PO $(cat^{-g}\ h)^{-1}$ for TS-1. But the catalytic activity decreased more quickly in the case of TS-1, probably because the methanol cause the solvolysis of PO to substances with higher boiling points such as propylene glycol (PG) and its MME. The by-products would cover the active sites to form coke and then lead to pore-blocking, which caused the deactivation of the catalyst more easily. In the case of Ti-MWW, the element analysis and the UV–visible spectrum of the fifth reused catalyst indicated that the nature of Ti sites did not change during the catalytic reactions [33]. After calcination, the activity was restored to the level of fresh

Table 4.2 Effect of solvent on liquid-phase propylene epoxidation over Ti-MWW[a]. Reprinted from Ref. [33], Copyright 2007, with permission from Elsevier

Solvent	n_{PO} (mmol)	PO selectivity[b] (%)	X_{H2O2} (%)	U_{H2O2} (%)
CH_3CN	23.9	99.9	83.0	94.0
Acetone	8.6	96.0	38.7	74.0
CH_3OH	5.1	96.0	35.8	48.0
1,2-Dichloroethane	8.2	97.4	38.6	71.0
DMF	0.8	47.8	5.28	45.4
THF	3.4	99.4	11.5	99.1
H_2O	4.2	98.0	26.4	52.4

[a] Reaction conditions: cat., 0.15 g; H_2O_2, 30 mmol; solvent, 10 g; pressure, 0.25 MPa; time, 1 h; temp., 313 K

[b] Propylene glycol (PG) and its monomethyl ethers (MME) were the main by-products

Fig. 4.1 The reuse of Ti-MWW and TS-1 in PO production. The used catalyst was regenerated by acetone washing and then drying at 353 K, except for No. 5 which was further calcined at 823 K. Reprinted from Ref. [33], Copyright 2007, with permission from Elsevier

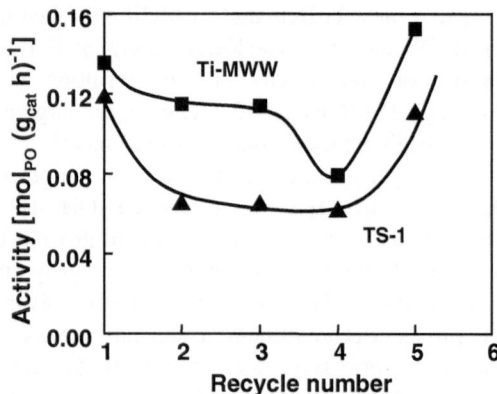

sample for both catalysts, indicating that the deactivation is due to the pore blocking by high boiling substance, which could be removed by calcination to regenerate the catalysts.

4.2.2 Unique Trans-Selectivity of Ti-MWW in Epoxidation of Cis/Trans-Alkenes with Hydrogen Peroxide

In the epoxidation of lower olefins, Clerici et al. reported that TS-1 was capable of producing epoxy derivatives with *cis*-configuration by starting from a mixture of alkenes containing *cis/trans* mixture of hex-2-enes [48]. The *cis*-selectivity of TS-1 is attributed to the higher reactivity of *cis* isomer in the epoxidation of various C_4 olefins. This result conflicts with the shape-selectivity because the *cis*-isomer with relatively larger molecular size is expected to be less active than the *trans*-isomer. However, in the case of Ti-MWW, the catalytic feature in the epoxidation of olefinic stereoisomers is totally opposite [27, 49]. Ti-MWW exhibits a singularity never observed on conventional titanosilicates, that it selectively epoxidizes the *trans*-isomer to give a selectivity of ca. 80 % for corresponding *trans*-epoxide from an alkene mixture with a *cis/trans* ratio of 50:50.

The epoxidation of hex-2-ene isomers with a *cis/trans* ratio of 41:59 was conducted over titanosilicates with various structures with hydrogen peroxide (Table 4.3). The products were 2,3-epoxyhexanes with both *cis-* and *trans*-configurations, and diols formed from a successive hydrolysis of epoxides over acid sites. A small amount of diol was obtained, especially for Ti-Beta, owing to the rather strong acidity relating to the framework Al. Both Ti-MCM-41 and amorphous SiO_2-TiO_2 exhibited high selectivity for diols due to the presence of silanol groups. Ti-Y, prepared by post-treating Y zeolite with $TiCl_4$ vapor, contained a large amount of silanol groups on the Al-deficient sites, thus also showed high selectivity for diols. These acidic silanol groups contributed to the hydrolysis of epoxides to diols. Ti-MWW showed the highest specific catalytic activity and higher efficiency in H_2O_2 utilization. It is interesting that Ti-MWW exhibited an extremely high selectivity for the epoxide with *trans*-configuration, while TS-1, TS-2, and Ti-MOR produced the *cis*-isomer more selectively.

The unique performance of Ti-MWW has been observed not only for hex-2-enes but also for other olefins such as 2-heptenes, 3-heptenes, and 2-octenes [27, 49]. Although the epoxidation activity decreased in the order of 2-heptenes, 3-heptenes, and 2-octenes, because of increasing steric constraint effect, the *trans*-selectivity for corresponding epoxide remained at high level. These phenomena further confirmed that Ti-MWW has a unique behavior in the epoxidation of *cis/trans*-alkene.

To confirm whether the unique *trans*-selectivity of Ti-MWW is due to a result of external factors, the epoxidation of hex-2-enes has been carried out in different solvents. Ti-MWW exhibited the highest catalytic activity in MeCN, but low

Table 4.3 Epoxidation of hex-2-ene isomer with hydrogen peroxide over various titanosilicates[a]. Reprinted from Ref. [27], Copyright 2002, with permission from American Chemical Society

Catalyst	Si/Ti	Conversion/ mol %	Selectivity/ mol %		Selectivity/ mol %		H_2O_2/ mol %	
			Epoxide	Diols	cis	trans	Conversion	Efficiency
Ti-MWW	46	50.8	99	1	98	1	55.1	92
TS-1	42	29.1	96	4	66	34	32.5	89
TS-2	95	13.6	96	4	67	33	18.0	77
Ti-Beta	40	15.9	91	9	73	27	35.8	45
Ti-MOR	79	2.6	99	1	52	48	3.9	66
Ti-Y	43	3.8	40	60	55	45	8.4	46
Ti-MCM-41	50	3.1	36	64	62	38	21.0	15
SiO_2-TiO_2	85	0.8	37	63	61	39	7.6	10

[a] Catalyst, 50 mg; hex-2-enes (*cis/trans* = 41:59), 10 mmol; H_2O_2, 10 mmol; MeCN, 10 mL; temp., 333 K; time, 2 h

activity in the protic solvent such as alcohols and H_2O. This behavior was opposite to that of TS-1, which prefers the protic solvents. Although the conversion rate for hex-2-enes varied in different solvents, the selectivity for *trans*-epoxide was slightly changed and maintained in the range of 73–83 % [27, 49].

Ti-MWW exhibited a unique selectivity for *trans*-configuration product, which was not observed on other titanosilicates in the epoxidation of *cis/trans*-alkenes with H_2O_2. To account for the reason, mechanistic investigation was carried out to clarify it [27]. Firstly, the liquid-phase adsorption of *cis/trans*-hex-2-ene was compared among Ti-MWW, TS-1, and Ti-Beta. TS-1 with medium pore showed higher adsorption capability than Ti-MWW and Ti-Beta owing to its hydrophobility framework, but adsorbed more selectively to the *trans*-isomer. Ti-MWW and Ti-Beta adsorbed two isomers almost nonselectively. If the diffusion was the rate-determining step in the epoxidation, TS-1 would show a higher activity for the *trans*-isomer to give a higher selectivity for the *trans*-epoxide than Ti-MWW and Ti-Beta. Therefore, the stereoselectivity of titanosilicates is hardly ascribed to the adsorption and diffusion of reactants.

It is necessary to confirm what reaction space of MWW zeolite is involved in the epoxidation of linear alkenes. The epoxidation of hex-2-enes was carried out in the presence of various poisons with different molecular sizes. Tripropylamine (TPA), which is able to enter both channels and open cages of MWW structure due to its small molecular size, reasonably poisoned all the Ti sites to deactivate Ti-MWW dramatically. On the other hand, bulky poisoning reagents of 2,4-dimethylquinoline (DMQ) and triphenylamine (TPhA) are considered to be too large to enter 10-MR channels of both MFI and MWW structures and selectively poison the active sites existing on external surface, and thus selectively block the Ti sites located in open space such as external cups. The results indicated that the DMQ and TPhA have little negative impact on the catalytic activity. Therefore, it can be concluded that the Ti sites within the cups on the external surface contribute little to the epoxidation of linear alkenes. However, it should be noted that Ti-MWW was lacking Ti species on the external surface due to serious acid treatment during preparation and thus showed negligible activity derived from the external cups.

The hypothesis was further confirmed by the epoxidation with the bulky oxidant TBHP (Table 4.4). TBHP would interact with Ti active sites to form the intermediate Ti alkylperoxo species (Ti-OOC(CH$_3$)$_3$). This bulky intermediate would retard the epoxidation due to significant steric constraint. As a result, both Ti-MWW and TS-1 showed low activity in the epoxidation of hex-2-enes, when using TBHP as an oxidant. Ti-Beta with three-dimensional 12-MR channels turned out to be the most active catalyst. However, it is interesting that the product selectivity greatly differed from what was observed when using H_2O_2. The product selectivity, obtained by TS-1, indicated that the Ti species on the external surface of TS-1 are nonstereoselective. Ti-Beta still remained its *cis*-selective behavior irrespective of the oxidant and even made this selectivity more pronounced. This was because the epoxidation by TBHP was still able to occur within the large

Table 4.4 Epoxidation of hex-2-ene isomers with TBHP over various titanosilicates[a]. Reprinted from Ref. [27], Copyright 2002, with permission from American Chemical Society

Catalyst	Catalyst amount/g	Si/Ti	Conversion/ mol %	Product selectivity/ mol %		Epoxide selectivity/ mol %		TBHP/mol %	
				Epoxides	Diols	cis	trans	Conversion	Efficiency
Ti-MWW	0.1	64	1.6	100	0	70	30	2.2	74
Ti-MWW	0.5	64	8.0	99	1	70	30	9.5	84
TS-1	0.5	42	3.1	99	1	47	53	3.6	85
Ti-Beta	0.1	40	14.4	98	2	93	7	16.7	86

[a] Hex-2-enes (*cis/trans* = 41:59), 10 mmol; TBHP (70 wt %), 10 mmol; MeCN, 10 ml; temp., 333 K; time, 3 h

channels characteristic of *cis*-selective nature. In the case of Ti-MWW, the use of TBHP changed Ti-MWW into a catalyst of *cis*-selectivity nature. As shown in Scheme 4.3, bulky TBHP restricted the reaction to the external surface, that is, mainly within the cups. Obviously, the *cis*-isomer having stereo-fitting with the cups would approach the Ti species, if located in the bottom of cups, more easily than the *trans*-isomer, resulting in high *cis*-selectivity. Therefore, the cups are considered to serve as reaction spaces to contribute the *cis*-selectivity.

The *trans*-selectivity of Ti-MWW is believed to originate from its 10-MR channels. The 10-MR system is defined by 2D sinusoidal channels restricted by an elliptical aperture of 4.0 * 5.9 Å [14]. The Ti peroxo species of five-membered cyclic structure may cause significant transition-state restriction especially within elliptical channels. The *trans*-alkenes would suffer less diffusion and hindrance than *cis*-alkenes, resulting in a higher epoxidation rate. As shown in Scheme 4.4, the special shape of 10-MR channels of MWW is more suitable to promote the epoxidation of *trans*-isomers. The channels of TS-1 and Ti-Beta are round in shape and relatively larger than the 10-MR channels of MWW structure in dimension. These channels are too large to impose steric hindrance on *cis/trans*-alkenes [27, 49].

Scheme 4.3 12-MR side cups of Ti-MWW favor the expodixation of *cis*-alkenes. Reprinted from Ref. [27], Copyright 2002, with permission from American Chemical Society

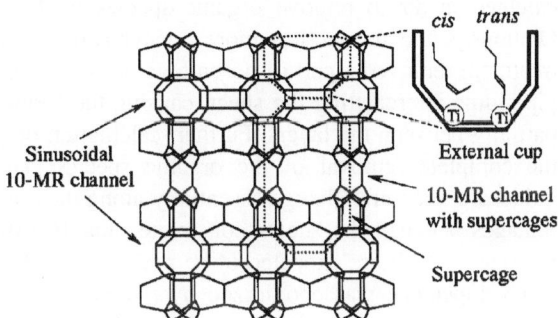

Sinusoidal 10-MR channel

cis trans

External cup

10-MR channel with supercages

Supercage

Scheme 4.4 Sinusoidal
10-MR channel system of
Ti-MWW favors the
epoxidation *trans*-alkenes

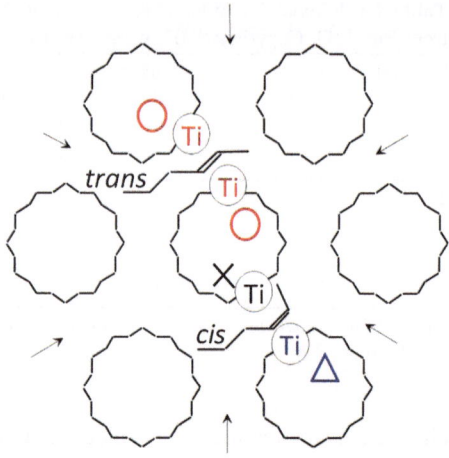

4.2.3 Selective Liquid-Phase Oxidation of Cyclopentene Over Ti-MWW

Owing to the unique structural properties, Ti-MWW has been proved to be active
and selective in the epoxidation of linear alkenes and also exhibits an extremely
high *trans*-stereo selectivity in the epoxidation of olefinic geometrical isomers.
Besides, the cycloalkene epoxides with bulky molecular size are also important
chemical intermediates in pharmaceutical and fragrant industry, such as cyclopentene epoxide. Therefore, the catalytic behavior of Ti-MWW in the epoxidation
of cycloalkene, such as cyclopentene, using H_2O_2 or TBHP as oxidant have been
checked and compared with other titanosilicates, such as TS-1 and Ti-Beta [30].

As we have known, the catalytic reaction involved larger substrate molecules
required more open reaction space. Thus, two kinds of Ti-MWW have been
prepared. The catalyst, prepared through acid treatment of Ti-MWW lamellar
precursor and drying, is denoted as Ti-MWW-dry. The sample that was further
calcined in air to remove organic species totally is denoted as Ti-MWW-cal.
Obviously, the former has more open reaction space than the latter. The acid
treatment removed a part of boron together with some structure directing agent of
piperidine intercalating the sheet, causing the formation of T-O-T linkage bonds
partially between the layers. Further, calcination of the acid-treated sample led to
the complete removal of the organic species. The interlayer dehydroxylation
induced a structural change essentially along the *c*-axis. The formation of ordered
linkages between the sheets constructed the 10-MR pores connected to the supercages, which are expected to be not so accessible as uncalcined samples.

Cyclopentene was epoxidized over three kinds of titanosilicates by H_2O_2 in
several representative solvents (Table 4.5). Besides cyclopentene oxide, the
products due to the solvolysis of epoxide and allylic oxidation were also observed.
Ti-MWW-dry was capable of giving cyclopentene conversion of 25.3 % and

Table 4.5 Epoxidation of cyclopentene over titanosilicates with H_2O_2 as an oxidant[a]. Reprinted from Ref. [30], Copyright 2006, with permission from Elsevier

Catalyst	Conversion/mol %	Selectivity/mol %				
		![epoxide] O	![OH] OH	![O] O	![OH/OH] OH/OH	![OR/OH] OR/OH
CH₃CN						
Ti-Beta	9.9	61.1	8.1	20.0	10.8	0
TS-1	13.3	91.8	1.9	5.3	1.0	0
Ti-MWW-dry	25.3	92.9	3.2	1.5	2.4	0
Acetone						
Ti-Beta	9.0	63.2	8.9	18.2	9.7	0
TS-1	10.3	72.6	8.8	17.0	1.6	0
Ti-MWW-dry	17.5	91.8	4.3	2.8	1.1	0
CH₃OH						
Ti-Beta	11.9	37.3	5.7	6.5	4.0	46.1
TS-1	16.1	70.6	2.8	3.3	2.6	20.7
Ti-MWW-dry	25.4	69.2	2.7	2.6	1.8	23.7
C₂H₅OH						
Ti-Beta	11.8	35.5	2.8	4.8	33.9	23.0
TS-1	14.3	52.8	12.9	6.3	20.3	7.7
Ti-MWW-dry	12.6	91.2	1.7	5.3	1.8	0

[a] Reaction conditions: cat., 50 mg; cyclopentene, 10 mmol; H_2O_2, 10 mmol; solvent, 10 mL; temp., 313 K; time, 2 h. The Si/Ti ratio was 35 for Ti-Beta, 48 for TS-1 and 34 for Ti-MWW-dry, respectively

cyclopentene oxide selectivity of 92.9 % in the aprotic solvent of MeCN, while it showed somewhat lower catalytic activity in the solvents of acetone and ethanol. Although comparable, high cyclopentene conversion was obtained in methanol for Ti-MWW-dry, low cyclopentene selectivity (69.2 %), due to the solvolysis of epoxide, was also observed. Since Ti-Beta containing a high portion of hydroxyl groups in the framework, the ring opening due to the acid-catalyzed hydrolysis of oxide occurred more easily independent of the solvents adopted. TS-1 showed a relatively high conversion for cyclopentene, but accelerated greatly the solvolysis of the oxide. Thus, Ti-MWW-dry has been proved to be a promising catalyst for the epoxidation of cyclopentene.

Nevertheless, it should be noted that the activity of Ti-MWW depends greatly on the post-treatment conditions, particularly the calcination. More open reaction space could be available in Ti-MWW-dry, which allows the substrate with relatively bulky molecular sizes reach the Ti active sites more easily. The difference between Ti-MWW-dry and Ti-MWW-cal was observed by conducting the cyclopentene epoxidation over a series of catalysts with various Ti contents. The conversion of cyclopentene increased reasonably with increasing Ti content for both kinds of Ti-MWW when the reactions were carried out with H_2O_2 under the same conditions. It was observed clearly that Ti-MWW-dry was superior to

Fig. 4.2 Liquid-phase adsorption of cyclopentene on Ti-MWW1-dry and Ti-MWW1-cal (*b*). The inset shows the change of adsorption amount at the initial stage. Adsorption conditions: adsorbent, 0.1 g; 2 wt % cyclopentene in 1,3,5-triisopropyl benzene, 2 g; temp., 273 K. Reprinted from Ref. [30], Copyright 2006, with permission from Elsevier

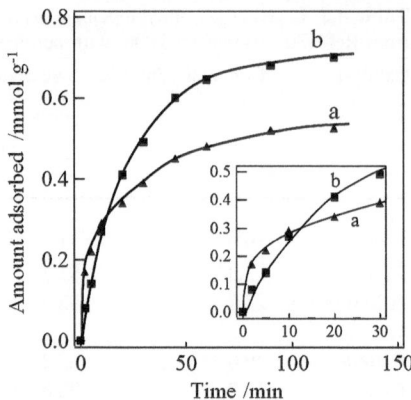

Ti-MWW-cal at the same level of Ti content. The selectivity to cyclopentene epoxide was generally higher than 90 % on both Ti-MWW catalysts.

The liquid-phase adsorption of cyclopentene was carried out in a solvent of TIPB, which is considered not to adsorb into 12-MR pores of zeolites because of a large molecular dimension. As the adsorption results in Fig. 4.2, Ti-MWW-dry adsorbed cyclopentene more rapidly than Ti-MWW-cal within a short time. But Ti-MWW-cal showed an adsorption capacity about 1.5 times that of Ti-MWW-dry after the cyclopentene adsorption amount leveled off. In the case of Ti-MWW-dry, the acid treatment would effectively remove the extra framework Ti species together with PI molecules intercalating the MWW sheets. Meanwhile, the interlayer entrance is kept open and penetrated easily by bulky molecules such as cyclic alkenes, which subsequently made the tetrahedral Ti active sites available to the oxidation reaction. As a result, Ti-MWW-dry with larger pore was superior to the calcined catalyst in the oxidation of cyclic alkenes.

Table 4.6 A comparison of cyclopentene epoxidation with H_2O_2 or TBHP over various titanosilicates[a]. Reprinted from Ref. [30], Copyright 2006, with permission from Elsevier

No.	Catalyst	Oxidant	Conversion/mol %	Selectivity/mol %		
				Epoxide	Diol	Allylic oxidation[b]
1	Ti-MWW1-dry	H_2O_2	24.5	92.9	2.4	4.7
2	Ti-MWW1-cal	H_2O_2	13.4	88.7	6.0	5.3
3	Ti-MWW1-dry	TBHP	23.4	92.9	0.1	7.0
4	Ti-MWW1-cal	TBHP	5.3	88.7	1.5	9.8
5	Ti-Beta	H_2O_2	9.9	61.1	10.8	28.1
6	Ti-Beta	TBHP	9.6	57.0	0	43.0
7	TS-1	H_2O_2	13.3	91.8	1.0	6.2
8	TS-1	TBHP	3.7	32.9	0	67.1

[a] Reaction conditions: cat. 50 mg; cyclopentene, 10 mmol; oxidant, 10 mmol; MeCN, 10 mL; temp., 313 K; time 2 h
[b] Products of allylic oxidation

The importance of the large pore of Ti-MWW-dry in the oxidation reaction involving large molecules has been further confirmed by the cyclopentene epoxidation with H_2O_2 or TBHP as an oxidant. In comparison to H_2O_2, TBHP with bulky molecular dimension greatly retarded the catalytic activity of Ti-MWW-cal, whereas Ti-MWW-dry showed comparable conversion for both oxidants (Table 4.6). The oxidant used has little effect on the catalytic performance of Ti-Beta, while the medium pore TS-1 was much less active with bulky TBHP as it proposed serious steric restriction.

4.2.4 Highly Efficient and Selective Production of Glycidol Through Epoxidation of Allyl Alcohol with H_2O_2

As discussed previously, taking advantages of this structural diversity, Ti-MWW has been proved to be active for the epoxidation of linear alkenes and cyclic alkenes as well. The unusual catalytic performance of Ti-MWW suggested that it may serve as a potential catalyst for the oxidation of functionalized alkenes.

Glycidol (GLY) is industrially produced by the epoxidation of AAL with H_2O_2, hypochlorous acid, or peracetic acid [50]. Among various oxidants, hydrogen peroxide is the most environmentally benign one, which requires to be catalytically activated. The conventional tungstate catalyst suffers from many problems such as activity, regeneration, and stability. TS-1 has been used as the catalyst in the epoxidation of AAL to GLY [48, 51–53]. However, the initial reaction rate for AAL epoxidation was reported to be about 30 times slower than the rate of 1-butene epoxidation. High selectivity to GLY was only observed when dilute H_2O_2 solution (5 wt %) was used. The low GLY selectivity was mainly due to the consecutive hydrolysis or solvolysis of intermediate GLY on the residual acid sites of TS-1. Moreover, the selectivity of GLY decreased from 100 to 2 % when the AAL conversion increased from 3 to 100 %, especially in the presence of alcoholic solvents of methanol or ethanol preferred by TS-1. Therefore, TS-1 cannot maintain a favorable balance between AAL conversion and GLY selectivity.

The catalytic properties of Ti-MWW in the epoxidation of AAL with H_2O_2 to GLY has been studied in detail. As shown in Scheme 4.5, the main product of the epoxidation of AAL is GLY, which may be hydrolyzed to glycerol through the ring opening reaction because aqueous H_2O_2 solution is used as an oxidant. The by-products of glycerol ethers may form due to the solvolysis of GLY when alcohols are used as solvent. Both the hydrolysis and solvolysis of GLY are due to the acidic character of titanosilicates derived from the silanol groups and the additional bridging hydroxyls of trivalent cations coexisting with Ti in the framework. Generally, the solvent has significant influence on the intrinsic activity of Ti species and product distribution in the AAL epoxidation.

Table 4.7 compares the catalytic performance of Ti-MWW with that of TS-1 and Ti-Beta for AAL epoxidation in various solvents. Ti-MWW showed highest AAL conversion, high GLY selectivity, and high efficiency for H_2O_2 utilization in

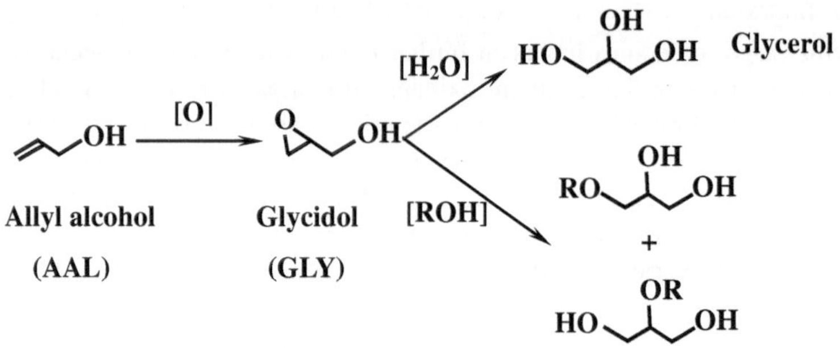

Scheme 4.5 Product distribution in AAL epoxidation. Reprinted from Ref. [28], Copyright 2003, with permission from Elsevier

MeCN and H_2O. The epoxidation of AAL was greatly retarded in the protic solvents of alcohols and also in the aprotic solvents of acetone and dioxane. As reported previously, the solvents alcohols play important roles in the formation of a stable five-membered cyclic intermediate species and stabilization of the Ti-hydroperoxo complex by the coordination of the protic molecule, ROH, to the Ti center through hydrogen bonding (Scheme 4.2). This species formed in the channels would gradually impose a steric hindrance on the substrates when ROH becomes bulky. Indeed, the AAL conversion slightly decreased in the order of MeOH > EtOH > 1-PrOH. Besides the influence on the catalytic activity, the protic alcohols caused the opening of the oxirane ring to greatly reduce the selectivity of GLY. On the other hand, TS-1 exhibited comparable AAL conversion in the protic solvents with relatively small molecular size such as H_2O and MeOH, and aprotic acetone. In contrast to Ti-MWW, TS-1 showed lower conversion in the epoxidation of AAL. It is obvious that TS-1 is much less effective in the epoxidation of AAL than Ti-MWW for both the activity and selectivity for GLY. In the case of Ti-Beta, very low AAL conversion was observed irrespective in protic solvents or aprotic solvents, and the GLY selectivity was much higher in MeCN than in the protic alcohols, probably because the basic MeCN may poison the Brønsted acid sites of Ti for the ring opening of epoxide. Considering that the Beta zeolite contains a concentration of acidic silanol groups too high at structural defect sites, lower activity of Ti-Beta should be ascribed to its hydrophilic character despite the Al-free form.

The extraordinarily high activity of Ti-MWW in the AAL epoxidation must be closely related to the unique crystalline structure of MWW topology. Selective poisoning of the Ti active sites with amines of different molecular sizes have been carried out to determine which part of Ti sites contribute more to the AAL epoxidation (Table 4.8) [28]. The poisoning reagents TPA and TPhA were used with different molecular sizes. The TPA with relatively small molecular size believed to reach all the Ti sites, and then deactivated Ti-MWW completely as expected. Nevertheless, TPhA with relatively large molecular size cannot go into the channels of MWW structure, and thus only deactivated the Ti active sites located at outer surface of Ti-MWW catalyst. The coexistence of TPhA had little

Table 4.7 Epoxidation of AAL with H_2O_2 in Various Solvents[a]. Reprinted from Ref. [28], Copyright 2003, with permission from Elsevier

Solvent	Ti-MWW (Si/Ti = 46)/mol %					TS-1 (Si/Ti = 36)/mol %					Ti-Beta (Si/Ti = 42)/mol %				
	AAL conversion	Product selectivity		H_2O_2		AAL conversion	Product selectivity		H_2O_2		AAL conversion	Product selectivity		H_2O_2	
		GLY	Others[b]	Conversion	Efficiency		GLY	Others[b]	Conversion	Efficiency		GLY	Others[b]	Conversion	Efficiency
MeCN	87.0	99.9	0.1	87.9	99.0	26.8	82.6	17.3	28.5	94.1	13.9	75.4	24.6	18.4	75.5
Water	82.3	99.9	0.1	84.3	97.6	34.6	96.0	4.0	36.6	94.5	2.8	92.6	7.4	9.6	29.2
MeOH	34.5	75.7	24.3	35.9	96.1	34.2	86.6	13.4	36.2	94.5	16.7	42.0	58.0	21.6	77.3
EtOH	32.5	91.0	9.0	33.0	98.5	24.4	94.6	5.4	29.8	81.8	15.1	59.5	40.5	28.6	52.8
1-PrOH	30.1	96.0	4.0	37.5	80.3	12.6	95.6	4.4	16.1	78.6	–	–	–	–	–
Acetone	41.5	96.7	3.3	42.5	97.6	31.0	92.8	7.2	36.6	84.7	11.9	41.4	58.6	26.3	45.2
Dioxane	27.8	96.0	4.0	28.6	97.2	–	–	–	–	–	5.2	78.3	21.7	6.5	80.0

[a] Cat., 70 mg; AAL, 10 mmol; H_2O_2, 10 mmol; solvent, 5 ml; temp., 333 K; time, 0.5 h

[b] Solvolysis products, glycerol and alkyl glycerol ethers etc

Table 4.8 Poisoning epoxidation of AAL over Ti-MWW[a]. Reprinted from Ref. [28], Copyright 2003, with permission from Elsevier

Poisoning reagent	Conversion/mol %	Product selectivity/mol %		H_2O_2/mol %	
		GLY	Glycerol	Conversion	Efficiency
None	92.6	99.5	0.5	93.7	98.8
TPA	1.3	87.7	12.3	–	–
TPhA	91.4	99.6	0.4	94.8	96.5

[a] Ti-MWW (Si/Ti = 38), 70 mg; AAL, 10 mmol; H_2O_2, 10 mmol; MeCN, 5 ml; amine, 2 mmol; temp., 333 K; time, 0.5 h

influence on the AAL conversion when compared to the nonpoisoning epoxidation. As Ti-MWW was prepared through severe acid washing toward the lamellar precursor, the acid treatment would selectively remove the Ti species on the external surface of the layers and then deeply decrease the Ti concentration within the side cups and supercages consisting of two cups [14]. Thus, it is assumed that the extreme high activity for the AAL epoxidation mainly originates from the intralayer 10-MR channels containing no supercages.

The reason of high GLY selectivity achieved on Ti-MWW should be taken into account. The outstanding oxidation activity may increase the epoxidation rate to a high level within a short time before significant solvolysis occurs. Alternatively, Ti-MWW may contribute little to the acid-catalyzed solvolysis. Concerning this issue, the hydrolysis of GLY has been carried out and the acidity of Ti-MWW, TS-1, Ti-Beta and Ti, Al-Beta has been characterized with NH_3-TPD [28]. Both TS-1 and Ti-MWW exhibited comparably low conversion of GLY to glycerol. The GLY conversion was slightly higher for Ti-Beta and much higher for Ti, Al-Beta. These results suggested that the weak acid sites in TS-1 and Ti-MWW had little influence on the GLY solvolysis. The degree of GLY solvolysis increased with the concentration of acid sites in the catalyst framework. The results of NH_3-TPD further confirmed the above results. Nearly defect-free TS-1 reasonably contained no acid sites, while Ti, Al-Beta had both weak and strong acid sites. These two kinds of acid sites were presumably attributed to the silanols at defect sites and framework Si(OH)Al groups, respectively. Pure silica Ti-Beta showed a low temperature peak, the same as the Ti, Al-Beta, with slight decrease in intensity. Ti-MWW contained some acid sites, weaker in strength and lesser in amount than pure silica Ti-Beta. The acid sites of Ti-MWW were mainly related to the framework B. Based on the above results, the acid sites of Ti-MWW are too weak to promote the hydrolysis or solvolysis of GLY effectively to yield corresponding glycerol and its derivatives. It is possible to completely remove the acidity related to B by repeating acid treatment on Ti-MWW, but it seems not necessary for AAL epoxidation.

4.2.5 Selective Epoxidation of Diallyl Ether with H_2O_2

The success in AAL epoxidation encouraged us to search for other applications of Ti-MWW in synthesizing oxygenated chemicals containing functional groups.

Allyl glycidyl ether (AGE) is an active monomer with high additional value which is useful as a diluting agent in synthesizing epoxy resins and alkyd resins, also used as an important stabilization agent for resins and agrichemicals [54]. AGE is mainly produced from the dehydrohalogenation between AAL and epichlorohydrin (ECII), which is synthesized from a stoichemical reaction of AAL and chlorine via the chlorohydrine. This process lacks greenness as it coproduces a large amount of unnecessary by-products. New synthesis methods based on heterogeneous catalysts are required to replace the old ones to improve the abovementioned drawbacks in the AGE synthesis.

Thus, Ti-MWW and conventional TS-1 catalyst were applied in the epoxidation of DAE in the presence of various solvents [29]. The epoxidation of DAE showed a relatively complicated product distribution as DAE contains two C=C bonds and underwent decomposition partially during the reaction, and furthermore the solvolysis reactions depending on the solvents used were also involved in the reaction. After identifying all the products by GC–MS and using authentic chemicals, the reaction pathways were clarified as depicted in Scheme 4.6. The main product of the DAE epoxidation is AGE, which can be successively oxidized to diglycidyl ether (DGE). Acid-catalyzed reactions also take place with AGE: (1) it is hydrolyzed at the ether group to form AAL and GLY; (2) its oxirane ring undergoes the solvolysis to produce ring-opening products when water (also existing in aqueous H_2O_2 solution), alcohol, and acetone are used as solvents. The reaction pathways of DAE firstly hydrolyzed to AAL and then oxidized to GLY are also possible. Theoretically, GLY can be further hydrolyzed to glycerol. Nevertheless, the amount of glycerol was negligible, because of trace concentration of GLY.

Table 4.9 compares the catalytic results of Ti-MWW with that of conventional TS-1 for the DAE epoxidation in various solvents. To compare the product selectivity at a similar DAE conversion, the catalytic reactions were carried out at different conditions for Ti-MWW and TS-1, since the former was much more

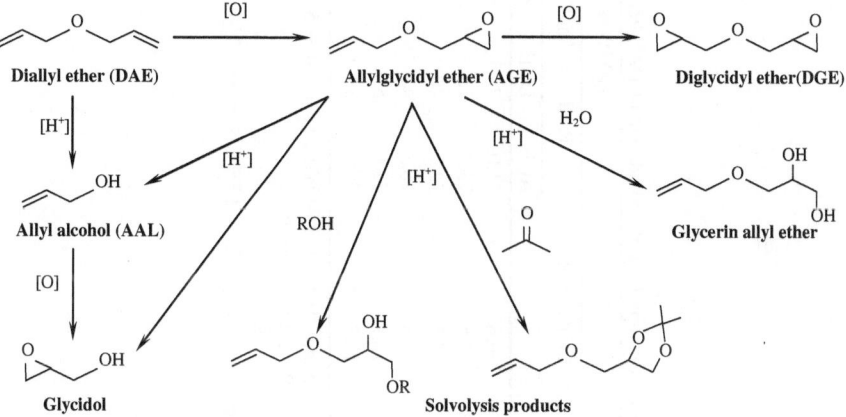

Scheme 4.6 Product distribution in DAE epoxidation. Reprinted from Ref. [29], Copyright 2004, with permission from Elsevier

Table 4.9 A comparison of epoxidation of DAE between Ti-MWW and TS-1[a]. Reprinted from Ref. [29], Copyright 2004, with permission from Elsevier

Solvent	Ti-MWW						TS-1					
	Conv. /mol%	Prod. sel. /mol%			H_2O_2 /mol%		Conv. /mol%	Prod. sel. /mol%			H_2O_2 /mol%	
		AGE	DGE	others[b]	conv.	eff.[c]		AGE	DGE	others[b]	conv.	eff.[c]
MeCN	39.6	71.0	28.7	0.4	99.8	95.2	19.1	77.7	15.4	6.9	42.8	48.6
Acetone	38.6	73.9	25.0	1.2	98.6	92.3	42.3	69.6	22.2	8.2	74.1	70.2
Water	25.3	52.7	40.0	7.3	78.5	86.5	6.6	60.6	13.5	25.9	38.2	18.5
MeOH	17.8	76.3	4.4	19.3	42.3	81.5	40.4	69.6	20.5	9.9	89.0	54.8
EtOH	12.1	79.8	5.7	14.5	38.7	60.5	33.9	71.7	15.4	12.9	76.7	50.2
Dioxane	18.7	79.2	19.0	1.8	46.7	93.7	23.3	88.6	6.2	5.2	44.9	52.3

[a] DAE, 10 mmol; H_2O_2:Ti-MWW(Si/Ti = 38) = 5 mmol:50 mg; H_2O_2:TS-1(Si/Ti = 36) = 10 mmol:200 mg; solvent, 5 mL; temp., 333 K; time, 0.5 h for Ti-MWW, 1.5 h for TS-1

[b] Solvolysis products etc

[c] H_2O_2 efficiency = (AGE + 2 DGE + others)/H_2O_2 converted * 100 %

active. Ti-MWW showed highest DAE conversion, similar AGE selectivity, and high H_2O_2 efficiency in MeCN and acetone, indicating both the solvents were suitable for Ti-MWW in the DAE epoxidation. The main by-product was DGE (25–29 %), while the amount of by-products derived from the solvolysis of AGE were very low. The phenomenon that the aprotic solvent of MeCN with weak basicity favored the oxidation of functionalized alkenes for Ti-MWW is similar to the solvent effect on the epoxidation of simple alkene as reported previously [26]. However, Ti-MWW showed low catalytic activity in the presence of protic solvents of water and alcohols and also in the aprotic solvent of dioxane. Water has been verified to be similarly efficient as MeCN for the epoxidation of AAL over Ti-MWW [28]. The difference between DAE and AAL is probably due to their different solubility in water, since the former is almost immiscible with water while the latter is totally soluble in water. The solvent effect involving water and alcohols was once assumed to be related to a stabilized five-membered cyclic intermediate species [48, 55]. The larger the dimension of these intermediate species has, the more serious the steric hindrance is. Thus, the DAE conversion decreased slightly in the order of $H_2O > MeOH > EtOH$. Besides, the protic solvents of alcohols would cause the solvolysis of AGE. In the case of DAE oxidation over Ti-MWW in MeOH, the by-products consisted of 80 % solvolysis product of AGE by MeOH, 15 % glycerin allyl ether, and 5 % AAL and GLY.

MeOH and acetone turned out to be the suitable solvents for TS-1 in the epoxidation of DAE, giving DAE conversion of around 40 %. However, both acetone and MeOH promoted the solvolysis of AGE, resulting in the oxirane ring-opening products over 8 %. MeCN, the most suitable solvent for Ti-MWW, showed somewhat lower conversion in the epoxidation of DAE. EtOH retarded the reactivity of TS-1, which was simply assumed to be the result of forming bulky and sterically restricted species as shown in Scheme 4.2. Besides, the utilization of H_2O_2 on TS-1 observed was lower in comparison with Ti-MWW. Considering that the epoxidation of DAE was carried out under much weaker reaction conditions for Ti-MWW, the catalytic results proved that Ti-MWW was obviously more effective than TS-1 in catalytic activity, epoxide selectivity, and H_2O_2 efficiency when choosing MeCN or acetone as a solvent.

The epoxidation of DAE consists of many reaction pathways, among which the consecutive oxidation of AGE to DGE is the main side reaction. From the viewpoint of producing AGE selectively, which is demanded in industrial and commercial applications, it is important to consider the issue of consecutive oxidation and propose some ideas for the selective production of AGE. To investigate the possibility of producing AGE selectively, the DAE oxidation was compared with the AGE oxidation using different amount of Ti-MWW [29]. It is clear that the reaction rate of DAE oxidation is much higher than that of AGE oxidation, probably because DAE with two C=C bonds have more chance to reach at Ti active sites in comparison with AGE. The other reason may be related to the hydrophilicity/hydrophobicity nature between DAE and AGE. DAE with more hydrophobicity may be easier than AGE to adsorb in the relatively hydrophilic Ti-MWW catalyst, and the subsequent epoxidation occurs quickly. This big

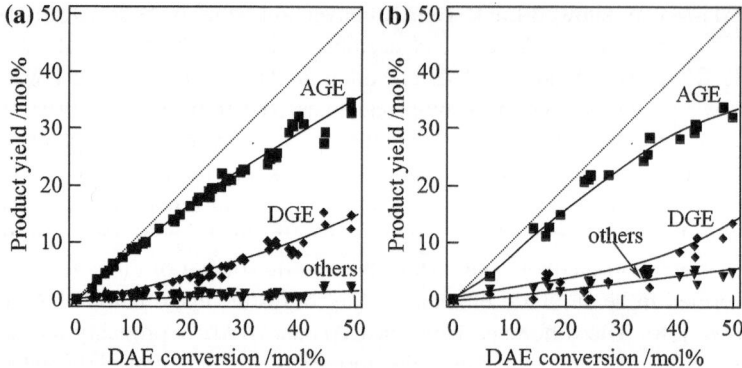

Fig. 4.3 Dependence of product yields on the DAE conversion on Ti-MWW (**a**) and TS-1 (**b**) at 333 K. Representative reaction conditions: 50 mg and 0.5 h for Ti-MWW; 200 mg and 1.5 h for TS-1. Reprinted from Ref. [29], Copyright 2004, with permission from Elsevier

difference in oxidation rate between DAE and AGE would make the selective production of AGE possible.

Figure 4.3 summarizes the dependence of product yield on the DAE conversion when the DAE epoxidations were carried out over Ti-MWW and TS-1. It should be noted that much more amount of catalyst and longer reaction time were needed in the DAE epoxidation over TS-1. The dotted line indicates the formation of AGE at a selectivity of 100 %. Both Ti-MWW and TS-1 showed similar profiles for the formation of AGE and DGE. The amount of AGE decreased with the reaction time, because of the consecutive oxidation of AGE to DGE. The results suggested that the high selectivity of AGE could be obtained, if the reaction rate was controlled at low level.

4.2.6 Epoxidation of 2,5-Dihydrofuran to 3,4-Epoxytetrahydrofuran

3,4-Epoxytetrahydrofuran (3,4-ETHF) is a useful intermediate in various chemical synthesis. It gives linear polymers with inherent viscosity and good hydrophilicity [56], serves as an important substrate for new desymmetrisation methodologies [57, 58], and provides a useful building block of tetrahydrofuran ring in pharmaceutical synthesis [59, 60]. At present, 3,4-ETHF is synthesized mainly through noncatalytic oxidation of 2,5-dihydrofuran (2,5-DHF) with peracids such as m-choloperbenzoic acid (CPBA) and trifluoroperacetic acid [61], and also using urea hydrogen peroxide together with phthalic anhydride [59]. All the methods suffer the disadvantages of production of a large quantity of toxic by-products, slow reaction rates, low product selectivity, and difficulty in product separation. The chlorohydrins method and an alternative way have also been applied to the synthesis of 3,4-ETHF, which are not environmentally benign because poisonous and corrosive halogens are involved in both the reactants and by-products. The TS-1/H$_2$O$_2$ catalytic system has been attempted in the epoxidation of 2,5-DHF [62]. Although it could give high conversion of 2,5-DHF, the 3,4-ETHF selectivity is somewhat low (70 80 %) due to the solvolysis of 3,4-ETHF in the solvent of methanol. Thus, Ti-MWW/H$_2$O$_2$ catalytic system was applied in the epoxidation of 2,5-DHF with hydrogen peroxide aqueous solution, attempting to solve the problems encountered in the conventional methods [32, 63].

As depicted in Scheme 4.7, the possible reaction pathways of the 2,5-DHF epoxidation could be classified into two parts: the first pathway is the oxidation taking place on the C=C bond to form 3,4-ETHF, which probably undergoes consecutive solvolysis in the presence of aprotic solvents such as alcohols and water. The acid sites of titanosilicates would contribute to the solvolysis of 3,4-ETHF to produce by-products. The acid sites including the silanols on defect sites, titanols, and Ti hydroperoxo species form as a result of the interaction of H$_2$O$_2$ with framework Ti. The second pathway is related to the reactions mainly occurring on the allylic position, in which both the oxidation and solvolysis are also involved. A direct bond formation between alcohol oxygen and carbon at

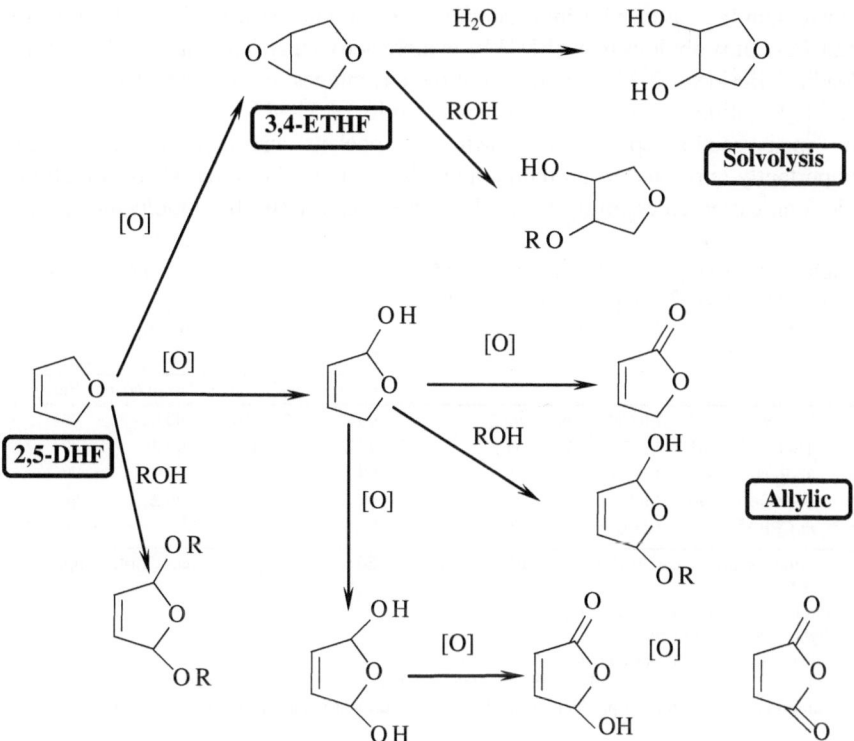

Scheme 4.7 Product distribution of 2,5-DHF epoxidation. Reprinted from Ref. [63], Copyright 2006, with permission from Royal Chemical Society

allylic positions of 2,5-DHF to corresponding ether by-products was also confirmed to occur. Therefore, the products of 2,5-DHF epoxidation could be sorted into three parts such as 3,4-ETHF, solvolysis products of 3,4-ETHF, and allylic oxidation products.

Table 4.10 compares the catalytic results of various titanosilicates with that of a homogeneous catalyst of m-choloperbenzoic acid (CPBA) for the 2,5-DHF epoxidation. The heterogenous reactions, catalyzed by titanosilicates, were carried out in the solvent of MeCN expect for TS-1, which used MeOH as solvent. Ti-MWW exhibited high 2,5-DHF conversion of 97 % and high 3,4-ETHF selectivity of 99 %. The catalytic activity based on TON decreased in the order of Ti-MWW > TS-1 > Ti-Beta > Ti-MOR. The homogeneous catalyst, m-CPBA, gave a 2,5-DHF conversion of 69.0 % and 3,4-ETHF selectivity of 97.4 %, which was much lower than those obtained with the heterogeneous Ti-MWW/H_2O_2 catalytic system.

The catalytic behavior of Ti-MWW in the 2,5-DHF epoxidation was further compared with that of TS-1 by applying a series of Ti-MWW and TS-1 with various Ti contents into the catalytic reactions. The Si/Ti molar ratios are in the range of 190–50 and 112–44 for Ti-MWW and TS-1, respectively. These catalysts were applied to the epoxidation of 2,5-DHF with H_2O_2 using the same amount of catalyst but in different suitable solvents. As shown in Fig. 4.4, the 2,5-DHF conversion increased with increasing amount of catalyst used for both Ti-MWW and TS-1. It is obvious that Ti-MWW is more active than TS-1 at the same catalyst loading. Besides, Ti-MWW possessed the advantages in the 3,4-ETHF selectivity and H_2O_2 utilization efficiency. The superiority of Ti-MWW may be attributed to its open reaction spaces of 12-MR side pockets and supercages, but more importantly attributed to its unique porosity such as sinusoidal shape which provides an easier accessibility to the Ti active sites for substrate molecules [14].

Table 4.10 A comparison of epoxidation of 2,5-DHF over various catalyts[a]. Reprinted from Ref. [32], Copyright 2007, with permission from Elsevier

No.	Catalyst	Si/ Ti	Solvent	TON[b]	Conversion (mol %)	Product selectivity (mol %)			H_2O_2 (mol %)	
						Oxide	Ally[c]	Others[d]	Conversion	Efficiency[e]
1	Ti-MWW	55	MeCN	229	97.2	99.8	0.2	0.0	99.8	98
2	TS-1	51	MeOH	156	71.3	74.0	14.8	11.2	99.3	72
3	Ti-Beta	42	MeCN	89	49.7	98.5	1.5	0.0	62.9	79
4	Ti-MOR	90	MeCN	67	17.4	99.5	0.5	0.0	20.2	86
5	m-CPBA[f]	–	CH_2Cl_2	–	69.0	97.4	2.6	0.0	–	–

[a] Reaction conditions: cat., 0.07 g; 2,5-DHF, 5 mmol; H_2O_2 (30 wt %), 5 mmol; solvent, 5 mL; temp., 333 K; time, 2 h

[b] Turnover number in mol (mol-Ti)$^{-1}$

[c] Allylic reaction products

[d] Solvolysis products of 3,4-ETHF

[e] H_2O_2 efficiency was calculated by relating all the oxidation products to the amount of H_2O_2 converted. The consumption of H_2O_2 for the products formed involving secondary and third oxidation has been taken into account

[f] The molar ratio of 2,5-DHF to m-CPBA was 1:1.9

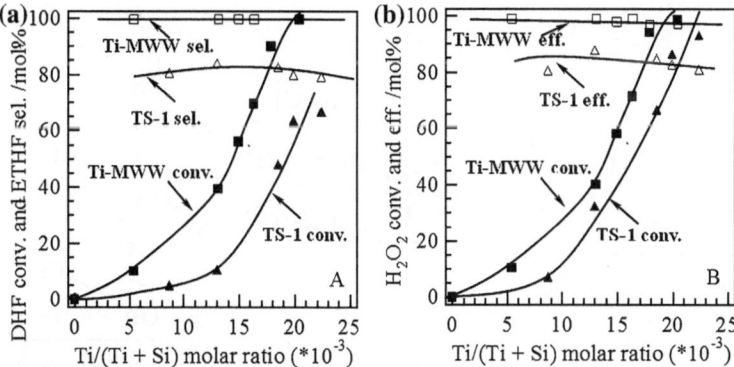

Fig. 4.4 Dependence of 2,5-DHF conversion and 3,4-ETHF selectivity (**a**), and H_2O_2 conversion and utilization efficiency (**b**) on the Ti content of Ti-MWW and TS-1. Reaction conditions: cat., 0.05 g; 2,5-DHF, 5 mmol; H_2O_2 (30 wt %), 5 mmol; solvent (MeCN for Ti-MWW and MeOH for TS-1), 5 mL; temp., 333 K; time, 2 h. Reprinted from Ref. [32], Copyright 2007, with permission from Elsevier

4.2.7 Selective Production of Epichlorohydrin Through Epoxidation of Allyl Chloride

Epichlorohydrin (ECH) is an important raw material for producing epoxy resins and synthetic glycerin. The conventional manufacturing methods used in current industrial processes are the high temperature chlorination of propylene and the method via allyl acetate. The key step of epoxidation in both processes depends on chlorohydrins and saponification, which inevitably produce a large quantity of calcium chloride by-product and halogen-containing wastewater. However, greener process has been developed based on TS-1/H_2O_2 system, which could direct epoxidation of ALC to ECH. But ECH is formed efficiently on TS-1 only when methanol used as solvent, which unavoidably leads to by-products due to the solvolysis of ECH [64, 65]. Therefore, Ti-MWW/H_2O_2 system was expected to produce ECH efficiently and selectively in the epoxidation of ALC [31].

As illustrated in Scheme 4.8, the main ALC epoxidation product was ECH. 3-Chloro-1,2-propanediol (CPDL) was co-produced due to the hydrolysis of ECH catalyzed by the weak acid sites such as silanols, titanols, or Ti peroxo species (Ti-OOH). ECH also could undergo solvolysis to produce hydroxyethers with high boiling points in the presence of protic solvents such as alcohols. Table 4.11 shows the results of ALC epoxidation with H_2O_2 aqueous solution over different titanosilicate catalysts. The reactions were carried out in solvent of acetonitrile expect for TS-1, which preferred methanol as solvent. Ti-MWW-PS, prepared by post-synthesized method, showed an ALC conversion of 83.4 % and an ECH selectivity of 99.9 % (Table 4.11, No. 1). Ti-MWW-HTS, prepared by hydrothermal synthesis, was slightly less active and selective than Ti-MWW-PS. The Brønsted acid

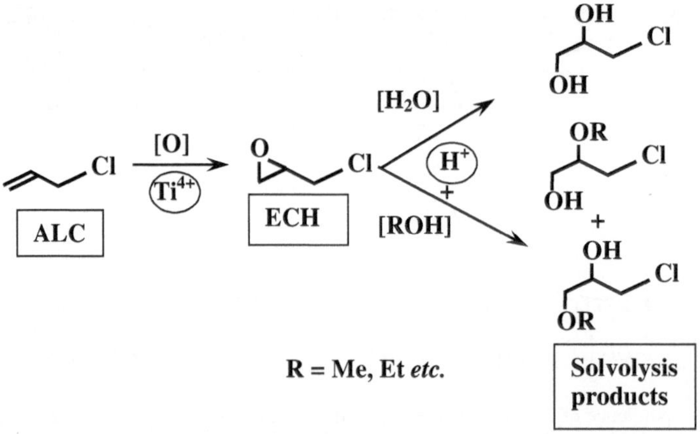

Scheme 4.8 Product distribution in ALC epoxidation. Reprinted from Ref. [31], Copyright 2007, with permission from Elsevier

Table 4.11 The result of epoxidation of ALC with H_2O_2 over different titanosilicate catalysts[a]. Reprinted from Ref. [31], Copyright 2007, with permission from Elsevier

No.	Catalyst	Si/Ti molar ratio	Solvent	ALC conversion (mol %)	ECH selectivity[b] (mol %)	TON[c]
1	Ti-MWW-PS	55	MeCN	83.4	99.9	275
2	Ti-MWW-HTS	37	MeCN	68.0	99.9	151
3	TS-1	47	MeOH	75.1	97.2	212
4	Ti-MOR	90	MeCN	1.6	98.6	9
5	Ti-Beta	76	MeCN	2.5	98.9	11

[a] Reaction conditions: cat., 0.1 g; allyl chloride, 10 mmol; H_2O_2 (30 wt %), 10 mmol; solvent, 5 mL; temp., 333 K; time, 2 h
[b] Others, mainly solvolysis products together with some heavy products
[c] *TON* turnover number in mol $(\text{mol-Ti})^{-1}$

sites originated from relatively high amount of framework boron seems not affect the ECH selectivity. The Ti species achieved by post incorporation are assumed to occupy the framework sites that are more accessible to the substrates than those obtained by direct hydrothermal synthesis, leading to a higher intrinsic activity (TON) for Ti-MWW-PS [66]. Under the same reaction conditions expect for solvent, TS-1 showed lower ALC conversion and ECH selectivity than Ti-MWW-PS (Table 4.11, No. 3). The superiority of Ti-MWW to TS-1 was attributed to the unique pore system of MWW zeolite favoring the adsorption and access of substrates to the Ti active sites [14]. Ti-Beta showed low ALC conversion and ECH selectivity due to its framework hydrophilic nature. Ti-MOR was also suitable for the epoxidation of ALC, because of its large crystals as well as one-dimensional channel. Therefore, despite the difference in the optimal reaction conditions, the titanosilicates exhibited catalytic activity for ALC epoxidation in the order of Ti-MWW-PS > TS-1 > Ti-MWW-HTS ≫ Ti-Beta > Ti-MOR.

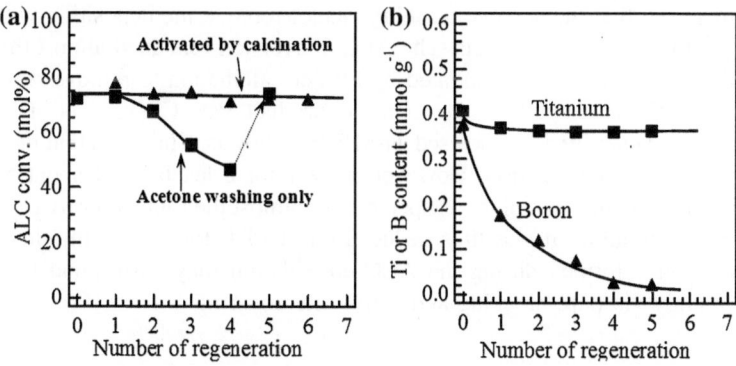

Fig. 4.5 Changes of ALC conversion (**a**), and Ti and B contents (**b**) with the reaction–regeneration cycles. Reaction conditions: Ti-MWW/ALC/H_2O_2/MeCN = 0.2 g/10 mmol/10 mmol/5 mL; temp., 333 K; time, 2 h. Regeneration: used catalyst was washed with acetone and dried at 393 K or further calcined at 773 K in air for 4 h. Reprinted from Ref. [31], Copyright 2007, with permission from Elsevier

The solvent effects have been investigated for Ti-MWW and TS-1 in the epoxidation of ALC [31]. The most favorable solvents for Ti-MWW were acetone and MeCN, in which higher conversion of ALC (>80 %), ECH selectivity (>99 %), and H_2O_2 utilization (>90 %) were obtained. As reported previously, the solvent effect of alcohols was once presumably related to a stabilized cyclic intermediate species of a five-membered ring [55]. It was actually observed that the ALC conversion decreased slightly in the order MeOH > EtOH > i-PrOH. Besides influencing catalytic activity, the protic alcohols caused the opening of the oxirane ring of ALC on the acid sites, which reduced the selectivity for ECH. The hydroxyl groups in the Ti-MWW framework may favor the adsorption of protic molecules such as alcohols and water, which retards the adsorption and coordination of substrates to the Ti active sites. Thus, Ti-MWW reasonably showed a higher conversion in a basic solvent or an aprotic solvent than in alcohols.

MeOH is the most suitable solvent for TS-1 in the epoxidation of ALC, giving an ALC conversion of 75.1 %. MeCN showed somewhat lower conversion for ALC. EtOH and i-PrOH also retarded TS-1 activity, due to the formation of bulky species with steric restrictions. However, TS-1 did not show a high selectivity for ECH in most suitable solvent of MeOH. The solvolysis of ECH with MeOH readily occurred to form CPDL and ethers. Furthermore, nonproductive decomposition of H_2O_2 was generally more likely to occur on TS-1, leading to a lower utilization efficiency of H_2O_2.

Ti-MWW was reused in the epoxidation of ALC to check the stability and reusability of catalyst. The used Ti-MWW catalyst was regenerated by washing with acetone or by further calcination in air at 773 K. The ALC conversion decreased dramatically with the reaction-regeneration cycles when the used catalyst was washed with acetone and then dried (Fig. 4.5). Nevertheless, the ALC conversion was restored after calcination at the fourth reaction-regeneration cycle.

This suggested that the acetone washing cannot remove the deposition with high boiling point inside the channels. The ALC conversion remained almost the same when the used catalyst was activated by further calcination to remove deposition totally. As discussed before, the main by-product was CPDL. The Ti and B contents were quantified for the used Ti-MWW catalysts. The B content decreased gradually during the reaction. However, expect for a leaching of about 5 wt % after the first use, the amount of Ti species was almost the same after five cycles of reaction-regeneration. It was thus deduced that CPDL together with other heavy organic species formed during the ALC epoxidation may correspond to deactivation mainly via pore blocking and active site covering.

4.3 Ammoximation of Ketones and Aldehydes

Ketone oxime or aldehyde oxime is an important fine chemical with a constantly increasing world market [67, 68]. The conventional manufacturing processes of oxime include the noncatalytic oximation of corresponding ketone or aldehyde with a hydroxylamine derivative like $(NH_2OH)_2 \cdot H_2SO_4$, and the separation of oxime by neutralizing sulfuric acid with ammonia. These conventional processes are encountering serious disadvantages such as using poisonous agents like hydroxylamine and corrosive sulfuric acid, producing valueless ammonium sulfate by-product, and bringing about environmental problems [67]. The titansilicate/H_2O_2 catalytic systems have found to be capable of catalyzing a variety of ketones or aldehydes with hydrogen peroxide and ammonia with water as almost the sole by-product [69]. The TS-1/H_2O_2 catalytic process has been commercialized in the liquid-phase ammoximation of cyclohexanone in 2003, which set a new milestone in zeolite catalysis.

Compared with TS-1 and other titanosilicates, Ti-MWW also exhibited efficient catalytic behaviors in the ammoximation of various ketones and aldehydes, giving ketones or aldehydes conversion and corresponding oxime selectivity over 99 % under optimized conditions, but Ti-MWW was particularly superior to TS-1 in oxime selectivity [34, 35].

4.3.1 Ammoximation of Methyl Ethyl Ketone

The liquid-phase ammoximation of methyl ethyl ketone (MEK) with ammonia and hydrogen peroxide was conducted on various titanosilicates such as Ti-MWW, TS-1, Ti-MOR, and Ti-Beta to produce methyl ethyl ketone oxime (MEKO) [35]. As the results listed in Table 4.12, the ammoximation of ketone almost did not proceed without catalyst, producing trace by-product due to aldol condensation. In the presence of titanosilicates, the 2-nitrobutane was the main by-product for the

Table 4.12 The results of liquid-phase ammoximation of ketones over different titanosilicates[a]. Reprinted from Ref. [35], Copyright 2007, with permission from Elsevier

No.	Catalyst[b]	Catalyst amount (g)	Solvent	MEK (mol %)		Cyclohexanone (mol %)	
				Conversion	Selectivity[c]	Conversion	Selectivity[d]
1	Without catalyst	–	H_2O	5.0	2.5	10.0	3.0
2	Ti-MWW(37)	1.5	H_2O	99.0	99.4	99.2	99.7
3	TS-1(35)[e]	2.0	H_2O/t-BuOH	99.0	80.0	97.3	99.8
4	TS-1(35)[e]	2.0	H_2O	20.0	95.0	18.0	78.8
5	Ti-MOR(90)	1.5	H_2O	72.0	98.3	65.0	96.5
6	Ti-Beta(76)	1.5	H_2O	25.5	85.0	15.0	20.0

[a] Reaction conditions: ketone, 100 mmol; H_2O_2, 120 mmol, NH_3 (25 wt %), 150 mmol; solvent, 25 g; temp., 335 K; time, 1.5 h. H_2O_2 (10 wt %) was added dropwise at a constant rate over 1 h
[b] The value in parentheses corresponds to the Si/Ti molar ratio
[c] The main by-product was 2-nitrobutane
[d] by-products were mainly peroxydicyckohexyl amine
[e] Reaction temp., 345 K

ammoximation of MEK. Ti-MOR and Ti-Beta were less active for ammoximation of MEK. Under the optimized reaction conditions, Ti-MWW is capable of showing a MEK conversion and a MEKO selectivity over 99 % when water was used as a solvent and dilute H_2O_2 was added dropwise into the reaction system. However, using a relative large amount of catalyst, the catalytic activity of TS-1 was significantly retarded by a sole solvent of water. In a co-solvent of t-BuOH and water, TS-1 showed comparably high conversion for MEK but lower selectivity of MEKO than Ti-MWW as results of co-producing 2-nitrobutane, which was almost absence in the case of Ti-MWW. A linear ketone like MEK with a smaller molecular dimension enters and diffuses freely into the pores of TS-1, which is helpful to the oxime formation. Nevertheless, the corresponding small molecule of MEKO formed inside the pores or diffusing into the channels would reach the Ti sites easily to induce a deep oxidation to 2-nitrobutane (Scheme 4.9). Intrinsically, Ti-MWW was an active catalyst for ammoximation but inactive for oxime formation. Therefore, both the MEK conversion and MEKO selectivity can achieve 99 % on Ti-MWW catalyst. Although the optimized conditions especially the solvent were different among the catalyst investigated, they showed a catalytic activity order of Ti-MWW > TS-1 > Ti-MOR ≫ Ti-Beta.

Scheme 4.9 Reaction pathways in MEK ammoximation. Reprinted from Ref. [35], Copyright 2007, with permission from Elsevier

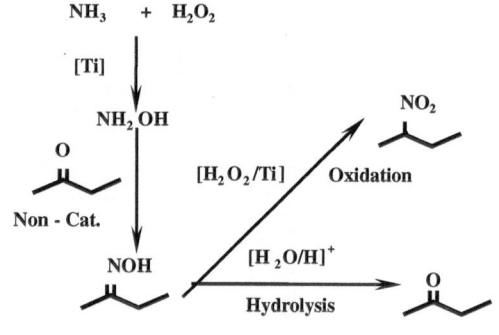

Fig. 4.6 The MEK conversion and the MEKO selectivity of the ammoxiamtion performed on a regenerated Ti-MWW. The used catalyst was regenerated by acetone washing and then drying at 393 K for 5 h (1–5th and 9th time), while it was activated by further calcination in air at 823 K for 6 h (6–8th time). Reprinted from Ref. [35], Copyright 2007, with permission from Elsevier

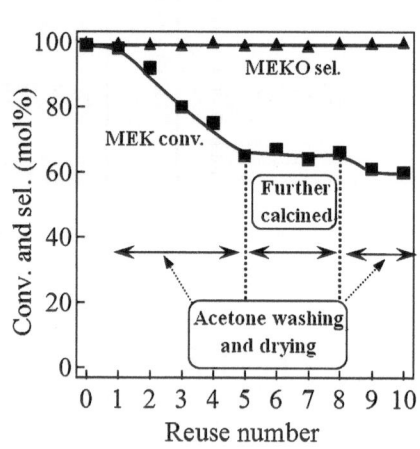

Besides its high catalytic activity and oxime selectivity, the reusability of Ti-MWW was important in the real applications. Thus, the stability and reusability of Ti-MWW in the MEK ammoximation have been checked [35]. After the used catalyst was washed with acetone and dried at 393 K, it was subjected to repeated ammoximation reactions at a constant ratio of catalyst-substrate-solvent (Fig. 4.6). The MEK conversion decreased to 65 % after fifth reuse, while the MEKO selectivity maintained at 99 % during five reaction-regeneration cycles. The catalyst gradually deactivated due to several reasons such as the deposition and pore blocking with heavy products, the leaching of active sites, the change of coordinated states of Ti species, and structural degradation of crystals. Thus, the fifth reused catalyst was regenerated by washing and drying followed by calcination in air at 823 K for 6 h. The regenerated catalyst showed almost unchanged MEK conversion and MEKO selectivity during the 6–8th reuse. However, the conversion dropped again when the used catalyst was only washed with acetone and dried at 393 K. These results showed us that the catalyst deactivated mainly due to pore jamming. High temperature calcination is helpful for prolonging the lifetime of Ti-MWW. However, it should be noted that the catalytic activity of Ti-MWW could not be restored completely by calcination, implying that there are other reasons governed the deactivation.

The used catalysts showed similar UV–visible spectra to the fresh Ti-MWW, indicating the change in the nature and coordination states of Ti active sites is not attributed to the deactivation [35]. The elemental analysis showed that the boron leaching continuously, which was often observed in the alkene epoxidation as we discussed before. On the other hand, the content of Ti species decreased by ca. 5 % after the first ammoximation reaction. However, the absolute Ti content per gram of catalyst increased slightly during further reuse, which was totally different from the phenomenon in the alkene epoxidation where the Ti content remained the same. These results were due to the partial dissolution of framework Si in the basic reaction media containing ammonia. The nitrogen adsorption further confirmed that the specific surface area decreased by ca. 20 % for the Ti-MWW catalyst reused six times. Thus, the desilication and partial structure degradation may account for the deactivation. The extraction of Si should create a high concentration of defect sites like hydroxyl nets in the framework. Ti-MWW with a greatly enhanced hydrophilicity would be disadvantageous to the selective oxidation in aqueous H_2O_2 solution. Considering the dissolution of framework Si is the main reason for deactivation, we added basic colloidal silica gel (30 wt %) directly into the reaction mixture and found that it was an effective method for prolonging catalyst lifetime [35]. The MEK ammoximation was carried out repeatedly with adding different amount of silica gel under optimized conditions. As shown in Fig. 4.7, the addition of a small amount of silica gel (6.7 wt % relative to MEK) decreased the deactivation rate and then prolonged the catalyst lifetime. Nevertheless, adding too much silica gel made the conversion decrease, which was due to the pore blocking and covering of Ti sites by silicon species. Thus, silica addition is expected to be served as a practical technique to improve the catalyst lifetime in the ammoximation of ketones.

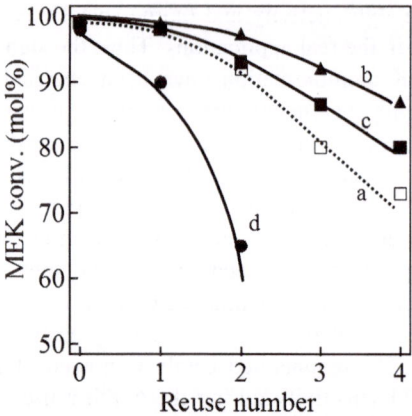

Fig. 4.7 Effect of colloidal silica gel addition on the MEK ammoximation on Ti-MWW. Reaction conditions: see Table 4.12 except that the amount is enlarged 20 times. The used catalyst was regenerated by acetone washing and then drying at 393 K for 5 h. The amounts of colloidal silica gel addition were 6.7 wt % (*a*), 13.4 wt % (*b*), 0 wt % (*c*), and (*d*) 67 wt % relative to MEK. Reprinted from Ref. [35], Copyright 2007, with permission from Elsevier

Fig. 4.8 The ketone conversion and the oxime selectivity of the ammoximation on structurally rearranged Ti-MWW. Ammoximation conditions: cat., 2wt %; temp., 338 K; H_2O_2/Ketone = 1.1:1; NH_3/Ketone = 2.2:1; solvent: *t*-BuOH/(*t*-BuOH + H_2O) = 50 %

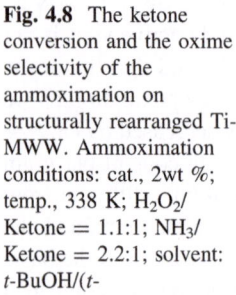
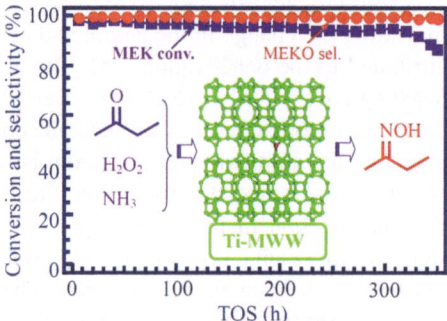

Based on the parameters of MEK ammoximation over Ti-MWW operated in batchwise, we have further developed a slurry process for producing MEKO continuously [70]. As shown in Fig. 4.8, structurally rearranged Ti-MWW showed stable activity and MEKO selectivity in a lab-scale test. At present, this clean process has been commercialized in China with a capacity of 15,000 t/y for MEKO production.

4.3.2 Ammoximation of Cyclohexanone

The catalytic performance of Ti-MWW in the liquid-phase ammoximation of cyclohexanone with ammonia and hydrogen peroxide was compared with other titanosilicates such as TS-1, Ti-MOR, and Ti-Beta [34]. The target product was cyclohexanone oxime. There were also some compounds with higher boiling points such as peroxydicyclohexyl amine. These by-products produced at a low cyclohexanone conversion, but their absolute amount was very low.

As the results given in Table 4.13, Ti-MWW-PS showed both cyclohexanone conversion and oxime selectivity over 99 % when H_2O_2 was charge into the reaction system slowly. Ti-MWW-HTS also showed high cyclohexanone conversion and oxime selectivity, when more catalyst was used. The Ti species achieved by post-incorporation were assumed to occupy the framework sites that were more accessible to the substrate molecules than those obtained by direct hydrothermal synthesis, leading to a higher intrinsic activity (TON) of Ti-MWW-PS [66]. In comparison with Ti-MWW, TS-1 showed comparably high cyclohexanone conversion and oxime selectivity, when increasing the amount of TS-1 relative to cyclohexanone and prolonging the reaction time. It should be noted that the activity of TS-1 was significantly retarded, when water was used as the sole solvent. Besides, Ti-MWW was obviously more active than TS-1 in terms of TON. This phenomenon was attributed to the unique pore system of MWW zeolite favoring the adsorption and access of substrate molecules to the Ti active sites. Ti-MOR and Ti-Beta were less active and selective. Therefore, although the optimized conditions especially the solvent were different among the catalyst investigated, they showed a catalytic activity order of Ti-MWW-PS > Ti-MWW-HTS > TS-1 > Ti-MOR > Ti-Beta.

The effects of the method of adding substrates on the cyclohexanone ammoximation also have been studied on some titanosilicates [34]. TS-1 showed somewhat improved cyclohexanone conversion when H_2O_2 or NH_3 was added into the reaction mixture dropwisely. Otherwise, this approach had no effect on Ti-MOR [12]. From the catalytic results given in Table 4.14, the adding method of the substrates has a significant influence on the catalytic behavior when Ti-MWW was used as catalyst. When a desired amount of H_2O_2 was added slowly and gradually into the reaction mixture of cyclohexanone, ammonia, water, and Ti-MWW, high cyclohexanone conversion and oxime selectivity were achieved. When all of the substrates and Ti-MWW catalyst were added at once, trace cyclohexanone oxime was obtained and the oxime selectivity was very low. When a desired amount of NH_3 solution was added dropwisely into the reaction mixture of cyclohexanone, H_2O_2, H_2O, and Ti-MWW catalyst, cyclohexanone conversion was still low, while the oxime selectivity was high. No matter how the method of adding substrates, the NH_3 conversion and H_2O_2 conversion were always high, indicating that the oxidation of NH_3 by H_2O_2 took place readily in all cases. The difference of the effect of adding substrates between Ti-MWW and other titanosilicates is presumed to be related to their catalytic properties and the ammoximation mechanism.

The catalytic activity of Ti-MWW greatly depends on the method of adding substrates, particularly H_2O_2. To make this issue clear, we have investigated the ammoximation mechanism [34]. The TS-1/H_2O_2 catalyzed the oxidation of an intermediate of cyclohexylimine formed by the noncatalytic reaction of cyclohexanone with ammonia was once considered to be the mechanism of ammoximation [71]. However, no obvious steric hindrance and diffusion problems were observed in the ammoximation of ketones and aldehydes with different molecular

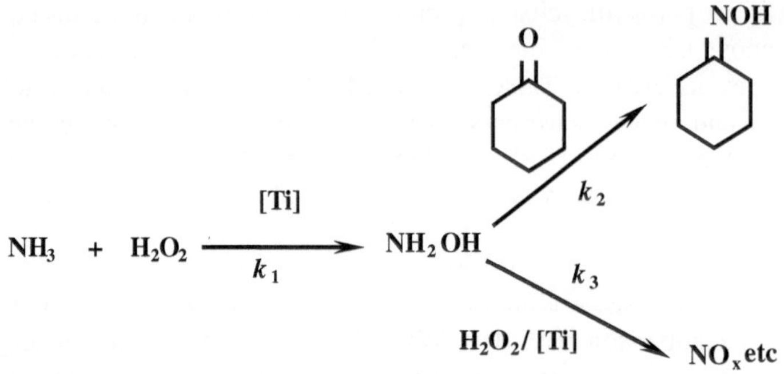

$$NH_3 \ + \ H_2O_2 \quad \xrightarrow[k_1]{[Ti]} \quad NH_2OH$$

Scheme 4.10 The reaction steps involved in the ammoximation. Reprinted from Ref. [34], Copyright 2006, with permission from Elsevier

Table 4.13 The results of cyclohexanone ammoximation over different titanosilicate catalysts[a]. Reprinted from Ref. [34], Copyright 2006, with permission from Elsevier

No.	Catalyst	Si/Ti	Solvent	Conversion (mol %)	Oxime selectivity[b] (mol %)	TON[e]
1	Ti-MWW-PS	55	H_2O	99.4	99.9	656
2	Ti-MWW-HTS[c]	50	H_2O	97.0	99.9	291
3	TS-1[d]	51	H_2O-t-BuOH	97.0	99.9	148
4	TS-1	51	H_2O	16.2	72.8	99
5	Ti-MOR	90	H_2O	60.0	95.0	122
6	Ti-Beta	76	H_2O	15.0	4.0	117

[a] Reaction conditions: cat., 50 mg; cyclohexanone, 10 mmol; solvent, 5 mL; NH_3 (25 %), 12 mmol; H_2O_2 (5 %), 12 mmol; temp., 338 K; time, 1.5 h. H_2O_2 was added dropwise at a constant rate within 1 h
[b] by-products were mainly peroxydicyclohexyl amine etc
[c] The amount of catalyst used was 100 mg
[d] The amount of TS-1 used was 200 mg. Reaction time was 5 h
[e] TON (Turnover number in mol (mol of Ti)$^{-1}$)

dimensions, indicating that the reaction may be through hydroxylamine route [72, 73]. As shown in Scheme 4.10, the ammoximation reaction consisted of the Ti-catalyzed hydroxylamine formation and the noncatalytic oximation of ketone with hydroxylamine to corresponding oxime. A detailed study of ammoximation over Ti-MOR has clearly verified this plausible reaction pathway [12]. Concerning the ammoximation mechanism related to hydroxylamine, we carried out the oxidation of NH_3 with H_2O_2 over Ti-MWW in the absence of cyclohexanone. After reaction, the Ti-MWW catalyst was removed by filtration, and an excessive amount of cyclohexanone was added to quantify the hydroxylamine formed. The catalytic results confirmed that the hydroxylamine intermediate actually formed and the amount of hydroxylamine intermediate obtained when adding H_2O_2 dropwisely was more than that obtained when adding H_2O_2 all together at the beginning. This result suggested that ammoximation over Ti-MWW also proceeded through the oxidation of ammonia by hydrogen peroxide to the

Table 4.14 Effect of adding method of substrates on the cyclohexanone ammoximation over Ti-MWW[a]. Reprinted from Ref. [34], Copyright 2006, with permission from Elsevier

No.	Conversion (mol %)	Oxime selectivity (mol %)	H_2O_2 conversion (mol %)	NH_3 conversion (mol %)
1[b]	99.4	99.9	98.4	96.0
2[c]	2.8	Trace	99.7	81.2
3[d]	4.8	95.0	99.8	84.3

[a] Reaction conditions: cat., Ti-MWW-PS (Si/Ti = 55); others, see Table 4.13
[b] A desirable amount of H_2O_2 was added dropwise for 1 h to the reaction mixture of cyclohexanone, NH_3, H_2O and Ti-MWW
[c] All the substrates were added once from the beginning
[d] A desirable amount of NH_3 solution was added dropwise for 1 h to the reaction mixture of cyclohexanone, H_2O_2, H_2O and Ti-MWW

intermediate hydroxylamine, followed by the oximation of ketone with hydroxylamine to oxime.

To further clarify the reason why oxime formation over Ti-MWW depended so greatly on the method of adding H_2O_2, we carried out the oxidation of hydroxylamine (in the form of hydroxylamine chloride) with H_2O_2 under the similar conditions over Ti-MWW and TS-1 [34]. Hydroxylamine was readily decomposed in the presence of H_2O_2 within a short time at a relatively low temperature. Moreover, Ti-MWW exhibited much higher oxidation activity for hydroxylamine than TS-1. Therefore, the concentration of free H_2O_2, that is H_2O_2 adding method, in the reaction mixture have great impact on the whole ammoximation reaction in the case of Ti-MWW catalyst. The nonproductive oxidation of hydroxylamine in the presence of free H_2O_2 on Ti sites to NO_x and other compounds also occurs during the ammoximation process. Hydroxylamine was produced rapidly, but the oxidation and decomposition of hydroxylamine became obvious and subsequently decreased the oxime yield, particularly when free H_2O_2 existed in the reaction mixture. This difference in the oxidation ability of hydroxylamine may be attributed to the different effects of the H_2O_2-adding method on Ti-MWW and TS-1.

4.3.3 Ammoximation of Cyclohexanone in a Continuous Slurry Reactor

Ti-MWW exhibited efficient catalytic properties in the ammoximation of ketones using a batchwise reactor [34]. These catalytic results implied that Ti-MWW is a promising catalyst for the clean synthesis of oximes. Nevertheless, the production of bulk chemicals like cyclohexanone oxime on a large scale favors a continuous slurry-bed reactor, which is equipped with a membrane separator to save the labor found in batchwise processes. To test the possibility of actual application of Ti-MWW catalyst, we applied Ti-MWW to the ammoximation of cyclohexanone in a continuous slurry reactor by simulating the operating conditions in a commercial processing procedure and studied the effects of reaction parameters, deactivation behavior and catalyst regeneration by comparing them with those from TS-1 under the same operation conditions [36].

The reaction temperature exhibited a great influence on the ammoximation of cyclohexanone over Ti-MWW in a continuous slurry reactor. The ammoxiamtion proceeded most effectively at an optimum reaction temperature of 343 K. When the temperature was further raised to 353 K, both the conversion and selectivity decreased, which was attributed to the faster vaporization and decomposition rates of reactants [36]. The ammoximation of cyclohexanone required stoichiometrically equivalent moles of NH_3 and cyclohexanone. Nevertheless, the cyclohexanone conversion and oxime selectivity reached ca. 96 and 99 %, when the NH_3/ cyclohexanone ratio and H_2O_2/cyclohexanone ratio was 1.7 and 1.1 (Figs. 4.9, 4.10), respectively. The reaction needs an excess amount of NH_3 to supplement the evaporated gaseous NH_3 and a slightly more H_2O_2 to proceed to its highest possible level. But too much excess of H_2O_2 in the reaction mixture would accelerate the deep oxidation of oxime to organic by-products, and also induce successive oxidation of hydroxylamine intermediates. The ammoximation reactions were performed by keeping the feeding rates of cyclohexanone, NH_3, H_2O_2, and solvent constant, but varying the amount of Ti-MWW catalyst used. A relative low concentration of Ti-MWW catalyst is sufficient to achieve high cyclohexanone conversion and oxime selectivity, but the lifetime of the reaction depended greatly on the catalyst loading. When the amount of catalyst changed from 1.5 to 3.2 g, the ammoximation lifetime was prolonged from 91 to 214 h [36].

In addition to high catalytic activity and oxime selectivity, the lifetime and reusability were also important to the real applicability. Therefore, Ti-MWW was applied to the ammoximation reaction in a continuous reactor compared with conventional TS-1 (Fig. 4.11). Both the conversion and oxime selectivity reached 96 and 99 %, respectively, during the ammoximation reaction, but Ti-MWW showed much longer lifetime than TS-1. Ti-MWW remained active for 214 h, whereas TS-1 deactivated at 120 h. Obviously, Ti-MWW was superior to TS-1 in terms of catalyst lifetime [36].

To clarify the main reasons for the deactivation of Ti-MWW during ammoximation, we have checked the used catalyst carefully [36]. The deactivated catalysts showed in UV–vis spectra the characteristic band at 220 nm due to the tetrahedral Ti species, indicating that most of the Ti species remained at tetrahedral sites in the zeolite framework [14]. However, there was a shoulder at 275 nm, which was absence in the fresh sample. This band became more obviously with the reaction time, particularly it was clearly observed for the deactivated catalyst. The band was generally attributed to the octahedral Ti species [14]. When the deactivated catalyst was calcined to remove any adsorbed inorganic and organic species, a spectrum nearly the same as that of the fresh catalyst was observed, but a weak shoulder was still present around 275 nm. FTIR spectrum of fresh Ti-MWW was compared with those of Ti-MWW used for different time [36]. The used catalysts all showed bands in the range of 2900–3000 cm^{-1}, assigned to the stretching of saturated C–H bonds, and became more intense with the reaction

Fig. 4.9 Effect of NH₃/ cyclohexanone ratio on continuous ammoximation of cyclohexanone on Ti-MWW. Reprinted from Ref. [36], Copyright 2011, with permission from Elsevier

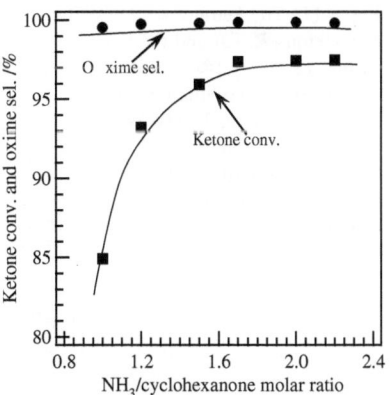

Fig. 4.10 Effect of H₂O₂/ cyclohexanone ratio on continuous ammoximation of cyclohexanone on Ti-MWW. Reprinted from Ref. [36], Copyright 2011, with permission from Elsevier

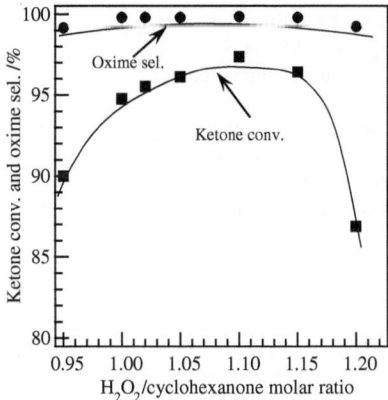

time. This revealed that some organic substrates with high boiling temperatures were deposited gradually inside the channels of Ti-MWW. These new bands disappeared completely when the used catalyst was calcined at 773 K in air, because the organic species were burned off. The N_2 adsorption investigation verified that both the surface area and pore volume of Ti-MWW decreased greatly during ammoximation, and were restored significantly by regeneration treatments. TG–DTA profiles were recorded to obtain the information concerning the coke deposition on Ti-MWW [36]. The weight loss increased with the reaction time, indicating the formation and deposition of organic by-products with high molecular weights on the surface or inside the channels of catalyst. These implied that the channels of Ti-MWW became severely blocked, and the Ti active sites were covered by by-product molecules.

As the catalyst was always immersed in basic media in the presence of excess NH_3, a partial dissolution of the silicalite framework was unavoidable. Therefore, decreased crystallinity also occurred in the ammoximation. The XRD patterns verified that the used catalyst still possessed the typical MWW structure, but its

Fig. 4.11 The ketone
conversion (■, □) and the
oxime selectivity (●, ○) of
continuous ammoximation of
cyclohexanone on Ti-MWW
and TS-1. Reprinted from
Ref. [36], Copyright 2011,
with permission from
Elsevier

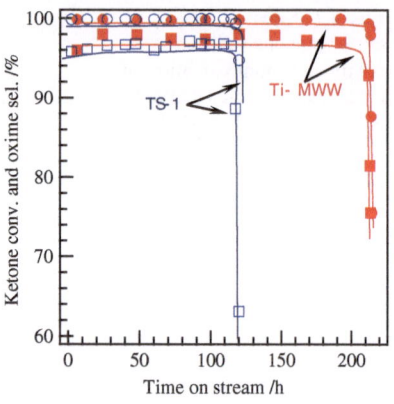

Fig. 4.12 Effect of silica
species addition on
continuous ammoximation
over Ti-MWW.
Ammoximation reaction
without (*a*) and with
(*b*) 5 ppm colloidal silica
addition. Reprinted from Ref.
[36], Copyright 2011, with
permission from Elsevier

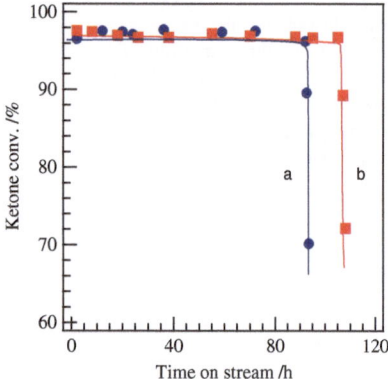

relative crystallinity decreased to a certain extent with prolonged reaction time
[36]. The apparent Si/Ti ratio of Ti-MWW decreased from 26 to 20 after been used
for 93 h, indicating that the content of Ti increased relatively to that of Si. The ^{29}Si
MAS NMR spectra of the deactivated Ti-MWW catalysts showed more Si(O-
Si)$_2$(OH)$_2$ (Q^2) and Si(OSi)$_3$OH (Q^3) groups, indicating that the desilication and
deboronation of the Si framework occurred during the ammoximation and defect
sites such as hydroxyl nests formed.

On the basis of the above results, the main reasons for catalyst deactivation
were assumed to be the coke deposition and desilication of catalyst framework. It
was possible to restrain desilication by controlling the silicon dynamic equilib-
rium. The colloid silica solution (30 wt %) with 5 ppm concentration was fed into
the reaction system [36]. Interestingly, the addition of silica gel postponed the
deactivation of Ti-MWW effectively, and prolonged the lifetime from 93 to 105 h
(Fig. 4.12). In contrast, these results suggested that the dissolution of the framework restricted the catalytic activity of Ti-MWW in a basic environment. The
method of regeneration was important for actual applications. The deactivated Ti-
MWW catalyst that was regenerated by calcination showed low activity for

cyclohexanone ammoximation, indicating that a simple calcination was insufficient for recovering the initial reactivity of Ti-MWW. The deactivated Ti-MWW catalyst after calcination was then subjected to structure rearrangement with PI followed by further calcination [74]. As we discussed before, this treatment would cause a reversible structural rearrangement for the MWW structure, partially mend the defect vacancies and enhance the hydrophobicity of the framework. The treated sample nearly restored the initial activity, but had a much shorter lifetime of 18 h [36]. Nevertheless, when the deactivated Ti-MWW catalyst was first treated with 0.3 M HNO_3 for 2 h at 353 K, then structurally rearranged with PI and calcined, the ammoximation lifetime was recovered effectively, showing a lifetime of 67 h [36]. The acid treatment in advance would remove selectively the extra-framework Ti species as well as organic substrates. The consequent structural rearrangement removed some of the defect sites. Thus, Ti-MWW is considered to be a promising catalyst for developing a much cleaner process of cyclohexanone ammoxmation.

4.3.4   Cyclohexanone Ammoximation with Core-Shell Structured Ti-MWW@meso-SiO$_2$

As we discussed before, the causes of catalyst deactivation of the cyclohexanone ammoximation in a continuous slurry reactor were elucidated to be the coke deposition and partial dissolution of the zeolite framework [36]. The ammoximation reactions prefer a basic environment created by an excess of NH_3. Therefore, a partial dissolution of the zeolite framework was unavoidable. In addition to the desilication, coke formation was another key factor accounting for catalytic deactivation. Based on these studies, core-shell-structured Ti-MWW@ meso-SiO$_2$ (MS) with a well-defined micro-meso hierarchical porosity was fabricated by using self-assembly technique and was applied as the catalyst for the ammoximation cyclohexanone in a continuous slurry reactor (Scheme 4.11) [37].

Figure 4.13 showed the HRTEM images of Ti-MWW@meso-SiO$_2$. Clearly, the materials comprised a core-shell structure with Ti-MWW crystal covered by mesoporous shell. Figure 4.14 compares the typical time course at the same loading of Ti-MWW active component, whereas the results using different catalyst weight are listed in Table 4.15. When the amount of Ti-MWW was varied from 1.0 to 2.0 g, the ammoximation lifetime was prolonged from 10 to 132 h. However, the MS-1.5TEOS catalyst exhibited a much longer lifetime than the parent Ti-MWW when compared at the same amount of the Ti-MWW active component. When the reactions were carried out using 2.0 g of active zeolite component, MS-1.5TEOS was deactivated at 198 h in comparison to 132 h for Ti-MWW. In the meanwhile, the physical mixture of Ti-MWW (70 wt %) and mesoporous silica (30 wt %), denoted as M&S, was also applied to the continuous ammoximation of cyclohexanone. However, the lifetime of M&S was only slightly longer than

Table 4.15 Textural and catalytic properties of Ti-MWW and Ti-MWW@meso-SiO$_2$. Reprinted from Ref. [37], Copyright 2013, with permission from American Chemical Society

No.	Catalyst	Amount[a] (g)	Lifetime (h)	Si/Ti[c]	SSA[d] ($m^2\ g^{-1}$)	PV[e] ($cm^3\ g^{-1}$)	Relative cryst.[f] (%)	Weight loss[g] (%)
1	Ti-MWW[b]	–	–	35	534	0.19	100	–
2	Ti-MWW	1.0	10	31	515	0.17	49.6	6.2
3	Ti-MWW	1.2	20	29	414	0.15	44.2	7.2
4	Ti-MWW	1.4	69	23	157	0.13	23.6	18.8
5	Ti-MWW	2.0	132	18	132	0.12	18.4	24.9
6	MS[b]	–	–	49	672	0.16	100	–
7	MS	1.43 (1.0)	34	43	301	0.10	69.2	9.7
8	MS	1.71 (1.2)	75	40	292	0.09	43.7	10.8
9	MS	2.00 (1.4)	127	36	287	0.02	36.1	12.1
10	MS	2.86 (2.0)	198	32	268	0.01	30.3	15.9

[a] The number in parentheses indicates the amount of Ti-MWW active component used
[b] Fresh sample
[c] Obtained by ICP analysis
[d] *SSA* specific surface area (BET)
[e] Micropore volume calculated by *t*-plot method
[f] Measured by accumulated intensity of the main XRD diffractions of [100], [101], [310]. The relative crystallinity of the fresh catalyst is assumed to be 100 %
[g] Obtained by TG measurement on deactivated catalysts without calcination

Ti-MWW, but much shorter than MS. This indicated that the added mesoporous silica could not protect the active component from deactivation as effective as the mesoporous silica shell does. In other words, the silica attached to the Ti-MWW surface could play the role in protecting the catalysts from deactivation much better than the additional silica source. Obviously, MS-1.5TEOS was more robust against deactivation and superior to Ti-MWW in terms of lifetime.

The causes of deactivation were elucidated to be the coke deposition and partial dissolution of the zeolite framework. The apparent Si/Ti ratio decreased from 35 to 18 after Ti-MWW was spent for 132 h, whereas that of MS-1.5TEOS decreased from 49 to 32 after used for 192 h. The increase of relative Ti content after reaction indicated that the desilication really occurred. The XRD patterns of their

Scheme 4.11 Strategy for prolonging ammoximation lifetime using core-shell catalyst. Reprinted from Ref. [37], Copyright 2013, with permission from American Chemical Society

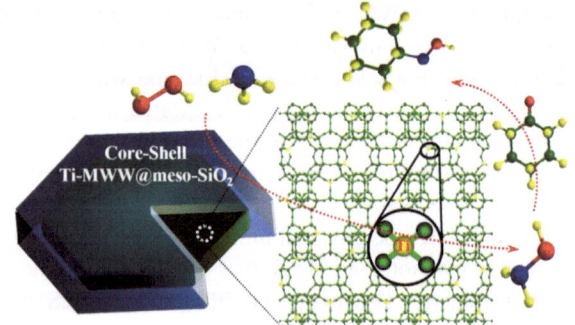

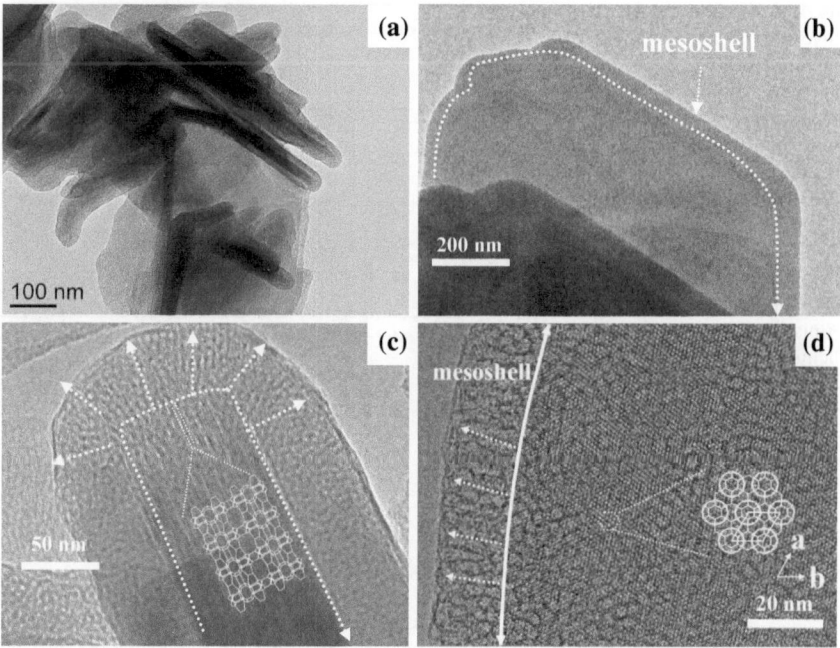

Fig. 4.13 HR-TEM images of Ti-MWW@mSiO$_2$-1.5TEOS taken along the directions of [100] (**a, c**) and [001] (**b, d**). Reprinted from Ref. [37], Copyright 2013, with permission from American Chemical Society

deactivated catalysts were recorded to investigate the desilication degree of the deactivated catalysts quantitatively [37]. Although the deactivated catalysts still possessed the typical MWW structure, their crystallinity decreased obviously. The relative crystallinity of Ti-MWW decreased to 18.4 % after being used for 132 h whereas that of MS-1.5TEOS decreased to 30.3 % at 198 h, indicating that the desilication degree of Ti-MWW was more serious than that of MS. This revealed that the mesosilica shell existing as an outer shell in MS may be dissolved in advance to Ti-MWW core. Serving as a sacrificial lamb, the shell protected the Ti-MWW crystallites from desilication effectively at least at the early stage of ammoximation. The protecting effect of mesosilica shell could be observed on the state and coordination of Ti active site in the deactivated and regenerated samples. The deactivated Ti-MWW showed the main adsorption band at 210 nm with shoulder bands at about 260 and 330 nm. These implied the deactivated sample contained the Ti species with six-coordination and in nonframework position. These adsorptions decreased in intensity to some extent, but did not disappear after the deactivated sample was regenerated. In contrast to Ti-MWW, the deactivated MS sample exhibited less obvious adsorptions at 260 or 330 nm. After regeneration, the spectrum was almost the same as the fresh MS-1.5TEOS sample.

Fig. 4.14 Cyclohexanone
conversion (**a**) and
cyclohexanone oxime
selectivity (**b**) in the
ammoximation on Ti-MWW
(*a*), Ti-MWW@meso-SiO$_2$
(*b*), and physical mixture of
Ti-MWW and mesosilica
(*c*) at the same loading of Ti-
MWW active component
(2.0 g). Reprinted from Ref.
[37], Copyright 2013, with
permission from American
Chemical Society

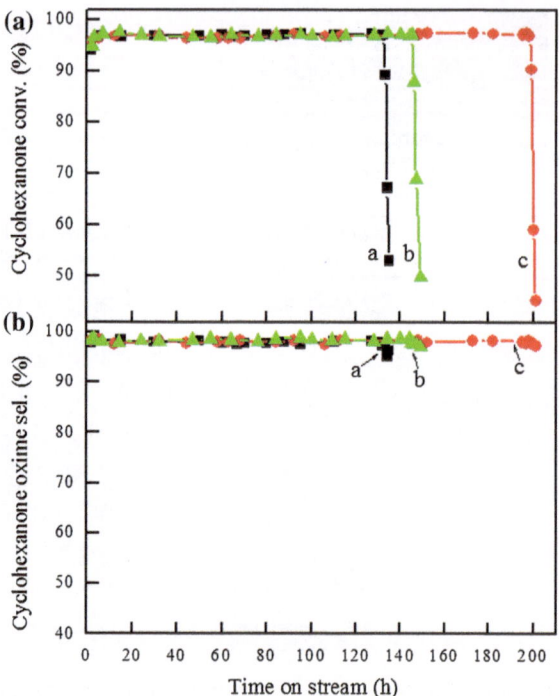

Therefore, the active Ti sites inside zeolite core got well protected by the presence
of mesosilica shell in the basic reaction media.

Besides the desilication, coke formation and deposition inside the zeolite
channels was another key factor accounting for catalytic deactivation. The high
boiling by-products were accumulated and deposited gradually inside the zeolite,
blocking the channels and covering the Ti sites severely. In comparison to the fresh
ones, the surface area and pore volume decreased greatly for both deactivated
Ti-MWW and MS-1.5TEOS without calcination [37]. TG–DTA profiles indicated
that the weight loss due to coke increased with the ammoximation lifetime, as a
result of gradual deposition of organic by-products. Interestingly, the coke deposition rate in deactivated MS increased more slowly in comparison to deactivated
Ti-MWW, while the absolute amount of coke deposition on MS was also less than
that on Ti-MWW. According to ammoximation mechanism, hydroxylamine was
in situ formed inside the titanosilicate pores first. This intermediate subsequently
diffused out of the crystallites pores and interacted with ketone to produce oxime
via a noncatalytic oximation route [34]. Because of the hydrophilic of the amorphous silica shell and the effect of spatially confinement, the existence of mesosilica
shell in MS-1.5TEOS may decrease the concentration of cyclohexanone at the pore
entrance of the core catalyst to certain extent. Thereafter, more coke may be formed
inside the mesosilica shell or outside the core-shell catalyst. With prolonging
processing time, the mesosilica shell was gradually dissolved into the reaction

solution. The coke deposited on the mesosilca shell then would be taken away simultaneously, which possibly lowered the coke deposition of Ti-MWW@meso-SiO$_2$ in comparison to Ti-MWW.

4.4 Unique Catalytic Properties of Fluorine-Planted Ti-MWW

MWW-type titanosilicate, Ti-MWW, has been proved to be a highly active catalyst for the liquid-phase oxidation of a variety of organic substrates using H$_2$O$_2$ as oxidant. Although excellent catalytic activity has been achieved, effective methods to further improve the catalytic activity are needed. As we discussed before, the Ti–O$^{\alpha}$–O$^{\beta}$–H peroxo species formed are considered to be the active intermediates (Scheme 4.12) [26]. In the case of alkene epoxidation, the C=C bond would capture the oxygen (O$^{\alpha}$) directly coordinated to the Ti sites to generate the corresponding epoxides, because this oxygen (O$^{\alpha}$) is more electrophilic than other ones (O$^{\beta}$). On the other hand, Ti-MWW possess little hydrophilic due to the silanols from the framework defects, which would have little negative effect on organic substrate adsorption and diffusion [14]. Thus, effective method was expected to be developed to improve the catalytic activity of Ti-MWW.

Fluorine has been implanted into a Ti-MWW framework through an acid treatment toward the Ti-MWW lamellar precursors in the presence of ammonium fluoride [75, 76]. Fluorine, the most electronegative element, would change easily the electropositivity of the elements in the neighborhood through an electron-withdrawing effect (Scheme 4.13). If the electropositivity around the tetrahedral Ti active sites is changed, the catalytic behavior of Ti-MWW would be improved further. Moreover, the fluorine modification could influence the solid acidity and hydrophilic/hydrophobic nature of zeolite by reacting with the terminal Si–OH groups to form Si–F species. The more hydrophobic nature of catalyst surface would favor the adsorption and diffusion of organic substrates.

Conventional Ti-MWW with a 3D MWW structure was prepared by treating hydrothermally synthesized lamellar precursors with acid-washing [14]. The F-implanted Ti-MWW, denoted as F–Ti-MWW, was prepared by adding NH$_4$F (Si/F = 26) into the above-mentioned acid treatment system for the precursors. The acid-treated Ti-MWW and F–Ti-MWW were calcined in air at 823 K to remove any residual organic species [75, 76].

The physical characterizations such as XRD, N$_2$ adsorption, UV–visible, verified that both the Ti-MWW and F–Ti-MWW are highly crystalline materials with tetrahedrally coordinated Ti ions in the framework [75, 76]. The surface Si/F molar ratio determined by the XPS data was 33, which was closer to the amount of F added in the acid treatment. A direct examination of fluorine environments in F-doped zeolites can be realized using ^{19}F NMR spectroscopy. Figure 4.15 shows the typical ^{19}F MAS NMR spectrum of F–Ti-MWW. It showed that two kinds of F species exist in F–Ti-MWW as Si–F and B–F interaction forms. The formation of the B–F species (-160.8 ppm) was due to the fact that the Ti-MWW precursor was synthesized using boric acid as a crystallization-supporting agent. The resonances at -129.1, -140.6, and -152.8 are attributed to the SiF_6^{2-}, $SiO_{2/4}F^-$, and $SiO_{2/3}F$ species, respectively. Consistent with the XPS measurement, the F species were incorporated into Ti-MWW by the NH_4F modification and the F^- ion which was favorably introduced consisted of the $SiO_{2/4}F^-$ and $SiO_{2/3}F$ units located in the framework.

The fluorine species that had a high electronegativity and existed as $SiO_{2/3}F$ units would withdraw the electrons of nearby Ti ions. As shown in Scheme 4.13, the Ti ions would become more positively charged, serving as stronger Lewis acid sites in F–Ti-MWW. In the actual catalytic oxidation reactions, the electrophilic interaction of these Ti sites with H_2O_2 molecules are presumed to be intensified, giving rise to more active Ti-peroxo species.

Table 4.16 compares between these two types of catalysts the results in the epoxidation of alkenes which differ in molecular structure and functional group. It is obvious that F–Ti-MWW showed an outstanding catalytic activity toward conventional Ti-MWW. As discussed before, when fluorine was implanted into the Ti-MWW framework, the surface hydrophilicity was also decreased. The improvement of framework hydrophilicity would favor the catalytic reaction. To clarify which aspect contributes the most to the improved catalytic activity of F–Ti-MWW, we first carried out the structural rearrangement with piperidine to

Scheme 4.12 Mechanism for alkene expodation catalyzed by titanosilicate/H_2O_2 system. Reprinted from Ref. [76], Copyright 2013, with permission from Royal Chemical Society

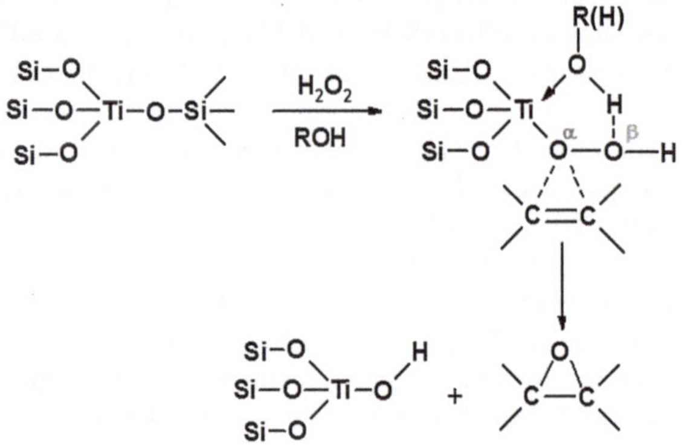

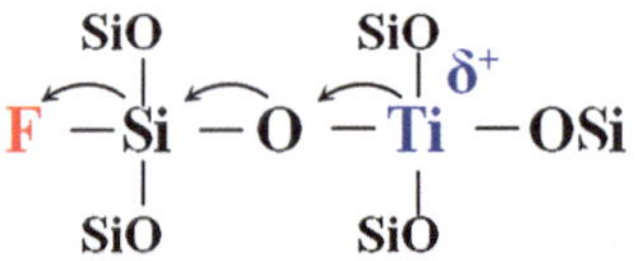

Scheme 4.13 Electropositive function of F in the F–Ti-MWW. Reprinted from Ref. [75], Copyright 2012, with permission from Royal Chemical Society

Fig. 4.15 ^{19}F MAS NMR
spectra of F–Ti-MWW
(a) and F–Ti-MWW-K (b).
The asterisks indicate the spin
side bands. Reprinted from
Ref. [75], Copyright 2012,
with permission from Royal
Chemical Society

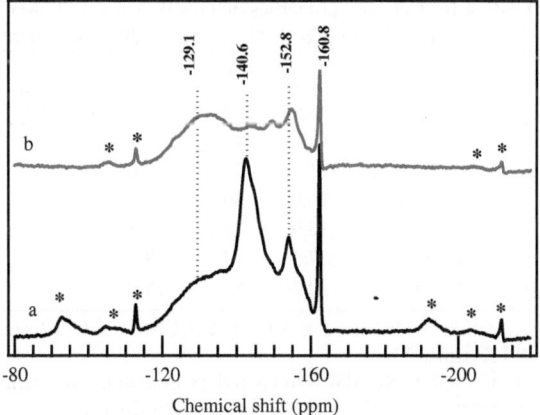

remove the hydrophilic silanols on defect sites, and then effectively enhanced the oxidation activity of Ti-MWW. However, the conversion of 1-hexene then increased from 41.6 to 47.2 %, far below the activity level of F–Ti-MWW (58.5 %). Presumably, the increased Lewis acid strength is a dominant factor governing the outstanding activity of F–Ti-MWW.

Ishihara has reported that F$^-$ could be removed through anion-exchange with aqueous sodium hydroxide. To eliminate the negative effect of the $SiO_{4/2}F^-$ units while keep the $Si_{2/3}F$ units intact, we have attempted the selective removal of the F species by employing a moderate KCl treatment. The ^{19}F NMR clearly evidenced that the $Si_{4/2}F^-$ units (−140.6 ppm) were remarkably reduced by the KCl treatment but the $Si_{3/2}F$ units (−152.8 ppm) were almost completely retained [75, 76]. The F–Ti-MWW-K catalyst showed a further improved epoxidation activity. The conversion of 1-hexene increased from 58.5 to 86.7 % with a well-preserved selectivity for the epoxidation product (Table 4.16, No. 6). The initial epoxidation rate of F–Ti-MWW-K became much higher, especially in high Ti content regions. The K treatment also increased the conversion of F–Ti-MWW in the epoxidation of other alkenes. In a control experiment, the same treatment made almost no difference to F-free Ti-MWW. Based on these results, it is deduced that there must be an interaction between F–Ti-MWW and KCl, which led to enhanced activity in F–Ti-MWW-K. Therefore, once the negative effect concerning the $Si_{4/2}F^-$ units was eliminated, the benefits of the remaining $SiO_{3/2}F$ units became outstanding in the catalytic reactions.

F–Ti-MWW was further applied in the ammoximation of cyclohexanone with NH$_3$ and H$_2$O$_2$ in a continuous slurry reactor by comparison with conventional Ti-MWW (Fig. 4.16) [76]. At the same weight hourly space velocity (WHSV), both cyclohexanone conversion and the cyclohexanone oxime selectivity were maintained at >96 and >99.5 %, respectively. The conventional Ti-MWW was deactivated in 210 h, whereas F–Ti-MWW remained active for 268 h. Thus, exhibiting a longer lifetime, F–Ti-MWW was more robust and stable than Ti-MWW in a continuous reaction.

Table 4.16 Catalytic properties of Ti-MWW and F–Ti-MWW in the epoxidation of alkenes with H_2O_2[a]. Reprinted from Ref. [75], Copyright 2012, with permission from Royal Chemical Society

No.	Substrate	Ti-MWW (%)			F–Ti-MWW (%)		
		X_{alkene}	$X_{H_2O_2}$	Oxide selectivity	X_{alkene}	$X_{H_2O_2}$	Oxide selectivity
1	1-hexene	41.6	49.4	98.6	58.5	67.4	98.9
2	Allyl chloride	21.5	24.5	99.6	28.7	31.2	99.8
3	1-heptene	26.0	46.7	98.1	34.0	46.6	97.2
4	Cyclopentene	15.9	17.1	67.1	17.4	19.2	81.7
5	1-hexene[b]	47.2	53.2	98.7	–	–	–
6	1-hexene[c]	42.2	55.1	98.6	86.7	94.7	98.8

[a] Reaction conditions: cat., 0.05 g; CH_3CN, 10 mL; 1-hexene 10 mmol, H_2O_2, 10 mmol; temp., 333 K; time, 2 h
[b] Catalyzed by Re-MWW prepared by piperidine treatment
[c] Catalyzed by F-Ti-MWW-K prepared by KCl treatment

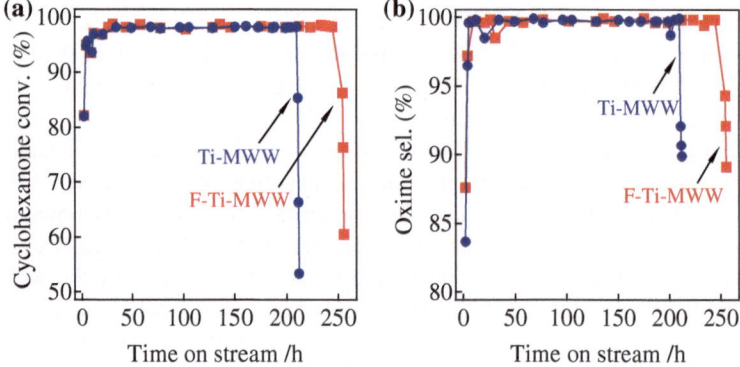

Fig. 4.16 The ketone conversion (**a**) and the oxime selectivity (**b**) of F–Ti-MWW (■) and Ti-MWW (●) in continuous ammoximation of cyclohexanone. Ammoximation conditions: catalyst, 3.2 g; 343 K; WHSV for cyclohexanone, 6.3 h^{-1}; H_2O_2/cyclohexanone = 1.1:1; NH_3/cyclo-hexanone = 1.7:1; solvent, t-BuOH (85 wt %). Reprinted from Ref. [76], Copyright 2013, with permission from Royal Chemical Society

4.5 Oxidation of Amines, 1,4-Dioxane, Sulfides and Olefinic Alcohols

4.5.1 Oxidation of Pyridines

Pyridine-N-oxide (PNO) is an important fine chemical with a constantly increasing world market owing to its usefulness as synthetic intermediates and biological importance. Heterocyclic PNO is also used as protecting groups, auxiliary agents, oxidants, ligands in metal complexes, and catalysts [77]. In addition, PNO is used

Table 4.17 The results of pyridine oxidation over different titanosilicate catalysts[a]. Reprinted from Ref. [38], Copyright 2010, with permission from Elsevier

No.	Catalyst[b]	Oxidant	Pyridine conversion (%)	PNO sel. (%)	X_{H2O2} (%)	U_{H2O2} (%)	TOF[c] (h^{-1})
1	Ti-MWW (43)	H_2O_2	97.7	99.4	96.0	77.8	252
2	TS-1 (40)	H_2O_2	99.2	95.5	97.1	75.1	238
3	Ti-Beta (70)	H_2O_2	51.5	95.7	70.2	54.0	216
4	Ti-MOR (92)	H_2O_2	24.5	90.2	60.3	28.2	135
5	Silicalite-1	H_2O_2	0	0	–	–	0
6	None	H_2O_2	0	0	–	–	0
7	Ti-MWW (43)	TBHP	13.0	99.0	–	–	33
8	TS-1 (40)	TBHP	1.4	98.0	–	–	3
9	Ti-Beta (70)	TBHP	2.4	97.6	–	–	10

[a] Reaction conditions: cat., 0.15 g; pyridine, 30 mmol; H_2O_2 (30 %), 39 mmol; temp., 348 K; time, 2 h; without solvent
[b] The number in parentheses indicates the Si/Ti molar ratio
[c] TOF: turnover frequency

as a functional chemical building block for synthesizing agrochemicals and pharmaceuticals. The conventional processes for manufacturing PNO are of multisteps involving sulfuric acid-catalyzed oxidation of pyridine with H_2O_2 as an oxidant, acetic acid as the solvent, and the separation of PNO by neutralizing sulfuric acid with sodium hydroxide [78]. These processes are encountering serious disadvantages such as using excessive solvent like acetic acid and corrosive sulfuric acid, producing onslaughts of valueless sodium sulfate by-product together with toxic waste solvent, and bringing about environmental problems.

The oxidation of pyridine to PNO with H_2O_2 or TBHP has been investigated on Ti-MWW and other titanosilicates [67]. There were no products obtained in the absence of catalyst or in the presence of Ti-free Silicalite-1, indicating that the isolated tetrahedral Ti^{4+} ions in titanosilicates are the active sites for pyridine oxidation. Superior to other titanosilicates like TS-1, Ti-Beta, and Ti-MOR, Ti-MWW showed a higher catalytic activity and PNO selectivity [38]. As the catalytic results in Table 4.17, Ti-MWW showed a pyridine conversion of 97.7 % and PNO selectivity of 99.4 % with H_2O_2 as an oxidant, while TS-1 showed a comparably high pyridine conversion but a lower PNO selectivity than Ti-MWW. The main products were 2-hydroxypyridine, 4-hydroxypyridine, oxaziranes, and azoxy compounds. However, the specific activity with respect to Ti species, turnover frequency (TOF), indicated that TS-1 was less active. The efficiency of H_2O_2 utilization was 75–78 % for Ti-MWW and TS-1. Nonproductive decomposition of H_2O_2 was mainly due to the basic media with the presence of pyridine. Ti-MOR and Ti-Beta turned out to be much less active and selective than Ti-MWW and TS-1. Since Ti-MOR had larger crystal size and only one-dimensional channels, both of which hinder the diffusion and access of substrate molecules to the Ti sites, making it not a suitable candidate for pyridine oxidation [12]. Due to highly hydrophilic features, Ti-Beta also showed low catalytic activity for pyridine

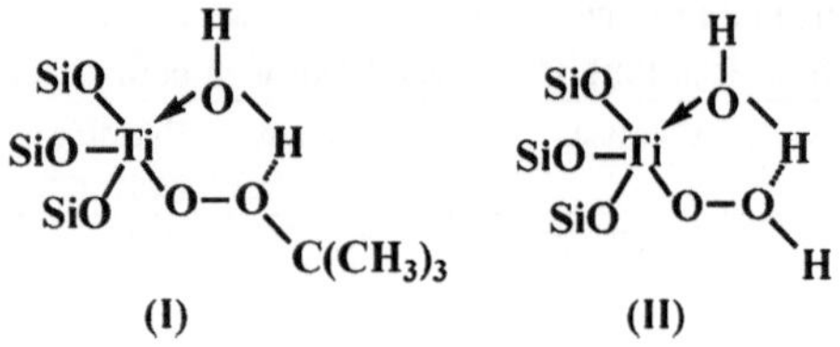

Scheme 4.14 Five-membered ring Ti species formed in TBHP and H$_2$O$_2$

(I)   (II)

oxidation. Therefore, in spite of the difference in optimum reaction conditions, the titanosilicates exhibited catalytic activity of pyridine oxidation in the order of Ti-MWW > TS-1 ≫ Ti-Beta > Ti-MOR.

In comparison with H$_2$O$_2$, the use of TBHP with a bulky molecular size made the reaction greatly retarded. In general, the titanosilicate-catalyzed reactions are considered to involve the 5-MR intermediates which are formed through the coordination of a solvent molecule such as alcohol or H$_2$O to a Ti peroxo species [26]. Species I and II shown in Scheme 4.14 are presumed to be the intermadiates for the oxidations with TBHP and H$_2$O$_2$, respectively, since both oxidants contained water. The species I is much larger than species II owing to a larger molecular dimension of *tert*-butyl groups. An actual reaction occurs only when the molecules can reach the above intermediates. The medium pores of TS-1 imposed a serious steric restriction for the intermediate of TBHP, which led to a very low activity. Similarly, the oxidation of pyridine with TBHP was restricted seriously on Ti-Beta. Although the catalytic performance of Ti-MWW was also affected by the molecular sizes of oxidant, it showed a higher pyridine conversion than TS-1 and Ti-Beta in the case of TBHP, which was attributed to the unique pore system of MWW zeolite with more open reaction spaces. It was these open reaction spaces that make Ti-MWW more active than TS-1 and Ti-Beta for the pyridine oxidation with TBHP.

The oxidation of pyridine derivatives with large molecular size have been performed on various titanosilicates (Table 4.18) [38]. The larger the substrate is, the more open reaction space is needed. The conversion of 3-picoline and 4-picoline was only 29.9 and 30.4 % over medium pore TS-1. Ti-MWW in the calcined form was more active than TS-1 by showing the conversion of 34.0 and 38.8 % for 3-picoline and 4-picoline, respectively. This is mainly because it has 12-MR supercages as well as external half cups. Nevertheless, the catalytic activity of Ti-MWW for these methyl-substituted derivatives was obviously lower than for pyridine as shown before. Ti-MWW-dry without calcination sample, which may still contained layered structure and then have open pore system, showed superior activity to the calcined Ti-MWW in the oxidation of pyridine derivatives. Inter-layer expanded MWW-type titanosilicate, IEZ-Ti-MWW, showed a much higher conversion than other catalysts [24]. The conversion of 3-picoline and 4-picoline reached 71.9 and 71.3 %, respectively. With the interlayer spacing expanded by silylation by 0.25 nm, IEZ-Ti-MWW possessed larger pore windows which were accessible to the substrates with large molecular dimensions and were also

Table 4.18 Oxidation of pyridine derivatives over different titanosilicates[a]. Reprinted from Ref. [38], Copyright 2010, with permission from Elsevier

No.	Catalyst	3-Picoline oxidation			4-Picoline oxidation		
		Conversion (%)	3-PNO selectivity[b] (%)	TOF[d] (h⁻¹)	Conversion (%)	4-PNO selectivity[c] (%)	TOF[d] (h⁻¹)
1	TS-1 (40)	29.9	97.0	72	30.4	96.5	73
2	Ti-MWW-cal. (43)	34.0	98.9	87	38.8	97.6	100
3	Ti-MWW-dry (43)	57.0	99.5	147	45.3	97.2	117
4	IEZ-Ti-MWW (40)	71.9	99.0	177	71.3	98.9	176

[a] Reaction conditions: cat., 0.15 g; 3-picoline or 4-picoline, 30 mmol; H_2O_2 (30 %), 39 mmol; temp., 348 K; time, 2 h; without solvent
[b] 3-PNO: 3-picoline-N-oxide
[c] 4-PNO: 4-picoline-N-oxide
[d] TOF: turnover frequency

beneficial for product desorption. As a result, the 3-PNO or 4-PNO yield decreased in the order of IEZ-Ti-MWW > Ti-MWW-dry > Ti-MWW-cal > TS-1.

4.5.2 Oxidation of 1,4-Dioxane

1,4-Dioxane is widely used for the synthesis of dyes, oils, paints, pesticides, and plastics [79]. 1,4-dioxane displays infinite solubility in water, however, it is probably a carcinogen and classified as a hazardous compound and a priority pollutant [80, 81]. Actually, it has contaminated groundwater and drink water in the USA and Japan. However, both the methods of bio-purification and adsorption by active carbon are not feasible. Fan et al. attempted to modify the chemical structure of 1,4-dioxane by means of redox reactions over titanosilicates like TS-1, Ti-Beta and Ti-MWW, and then significantly enhance its bio-degradability [82]. Ti-MWW exhibited high activity for the oxidation of 1,4-dioxane in the presence of H_2O_2 aqueous solution.

As we discussed before, solvent effect is an important issue in Ti-MWW-catalyzed reactions. Ti-MWW showed the best catalytic performance in the epoxidation of alkenes in acetonitrile [26]. The oxidation of 1,4-dioxane was performed in different solvents such as acetonitrile, methanol, acetone, and water, as well as performed without solvent (Fig. 4.17). 1,4-Dioxane-2-ol was the main product in all the reactions either with or without solvent. However, Ti-MWW exhibited low conversion of 1,4-dioxane in the oxidation reaction of 1,4-dioxane using acetonitrile as solvent. Moderate conversion of 1,4-dioxane was obtained when the catalytic reaction was performed in acetone or water. Interestingly, when the catalytic reaction was carried out in the absence of solvent expect for small amount of water present in the aqueous H_2O_2 solution, the conversion of 1,4-dioxane reached at 44.5 %. These results indicated that the presence of solvents

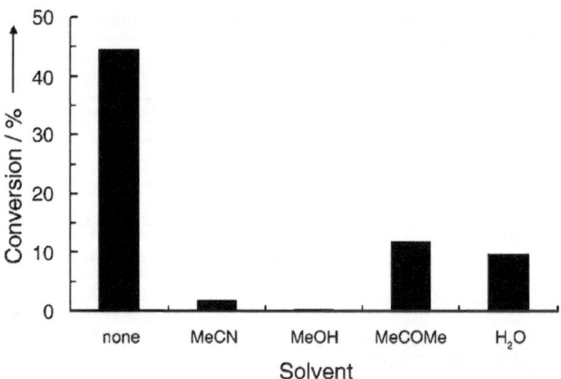

Fig. 4.17 Effect of solvent on the oxidation of 1,4-dioxane on Ti-MWW (reaction conditions in the absence of solvent: 0.1 g catalyst, 20 mmol substrate, 10 mmol H_2O_2 (31 % in aqueous solution), 60 °C, 4.4 h; in the presence of a solvent: 0.1 g catalyst, 10 mL solvent, 20 mmol substrate, 20 mmol H_2O_2 (31 % in aqueous solution), 60 °C, 4.4 h). Conversion based on H_2O_2. Reprinted from Ref. [82], Copyright 2008, with permission from John Wiley and Sons

retarded the catalytic ability of Ti-MWW in the oxidation of 1,4-dioxane to 1,4-dioxane-2-ol.

On the other hand, these results implied that the mechanism of 1,4-dioxane oxidation in the absence of solvents is different from those of the epoxidation of alkenes with acetonitrile as solvent. The intermediate with a five-membered ring structure formed by hydrogen-bonded solvent ROH or H_2O molecules and hydroperoxo-Ti species was considered to be the active site. It has been confirmed that oxidation of 1,4-dioxane with H_2O_2/O_2 occurs through a radical mechanism [83]. Firstly, hydrogen peroxide is decomposed homolytically to form a highly active HO· radical, which initializes the reaction. This process occurs slowly, as evidenced by a low H_2O_2-based conversion of 0.4 % obtained in the absence of catalyst. Then, this radical species abstracts one hydrogen atom from 1,4-dioxane to form a carbon radical. This is followed by its interaction with hydroperoxo/peroxo-Ti species to form radical Ti species, which further interact with hydrogen peroxide to produce the active intermediates of hydroperoxo-Ti species and hydroxyl radicals. The presence of HO· radicals was confirmed by a severe inhibition of the reaction in the presence of $Na_2CO_3 \cdot CO_3^{2-}$ is well known to be a scavenger of HO· radicals as shown. When Na_2CO_3 was added to the reaction system, the H_2O_2-based 1,4-dioxane conversion decreased from 46 to 3.6 %. Different amount of NaCl was also added into the reaction mixture, but the conversion of 1,4-dioxane remained at 30 %. The much higher activity obtained in the presence of NaCl than Na_2CO_3 showed that hydroxyl radical was really existed.

The catalytic ability of Ti-MWW in the oxidation of 1,4-dioxane was compared with other titanosilicates like TS-1, Ti-Beta. Ti-MWW showed much higher conversion than TS-1 and Ti-Beta [82].

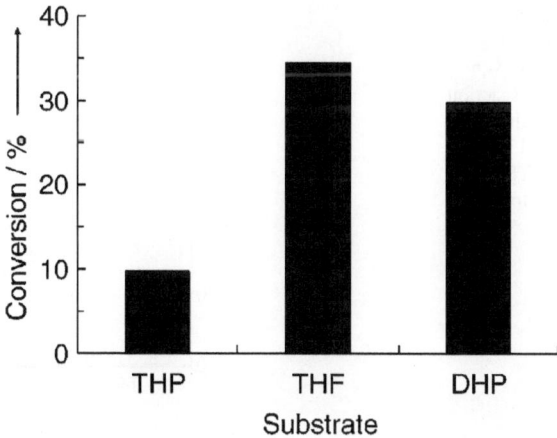

Fig. 4.18 Conversion of a series of ethers over the Ti-MWW catalyst (reaction conditions: for THP: 100 mmol substrate, 10 mmol H_2O_2 (31 % in aqueous solution) 60 $^\circ$C, 4.4 h, no solvent; for THF and DHP: 10 mmol substrate, 10 mmol H_2O_2 (31 % in aqueous solution), 10 mL MeCN, 60 $^\circ$C, 4.4 h). Conversion based on H_2O_2. Reprinted from Ref. [82], Copyright 2008, with permission from John Wiley and Sons

The catalytic oxidation of various ether substrates was performed over Ti-MWW catalyst (Fig. 4.18) [82]. In the case of tetrahydropyran (THP) oxidation, the conversion observed is 9.0 %, much lower than the conversion of 1,4-dioxane. The difference in substrates conversion was believed to be related to different solubility of these two substrates in water. Thus, a much higher concentration of 1,4-dioxane molecules than THP would be present around active sites. Acetonitrile is used as solvent in the oxidation of tetrahydrofuran (THF) and 3,4-dihydro-2H-pyran (DHP). Ti-MWW showed more than 30 % conversion for both THP and DHP. The results indicated that Ti-MWW could catalyze the oxidation of both saturated and unsaturated ethers under suitable conditions.

4.5.3 Oxidative Desulfurization with Ti-MWW

In recent years, environment problems is becoming serious, and thus strict regulations were implemented, that limited the level of sulfur in diesel to less than 15 μg/g since 2006 in the USA, less than 10 μg/g since 2005 in Europe, and less than 50 μg/g since 2008 in Beijing and Shanghai in China. Hydrodesulfurization (HDS) is efficient for the removal of thiols, sulfides, and disulfides. However, it is difficult to remove sulfur-containing compounds such as dibenzothiophene to reach a low level using conventional HDS. Therefore, the application of oxidative desulfurization to liquid fuels has attracted much interest. In this useful alternative, the sulfur compounds are first converted to oxygenated compounds with a high

Fig. 4.19 Conversion of benzothiophene (**a**) and dibenzothiophene (**b**) over TS-1 and Ti-MWW. Reaction conditions: model light oil 10 mL, 30 % H_2O_2 136 µL, $n(H_2O_2)$: $n(S)$ = 4, TS-1 or Ti-MWW 0.1 g, acetonitrile 10 mL, 343 K, 3 h

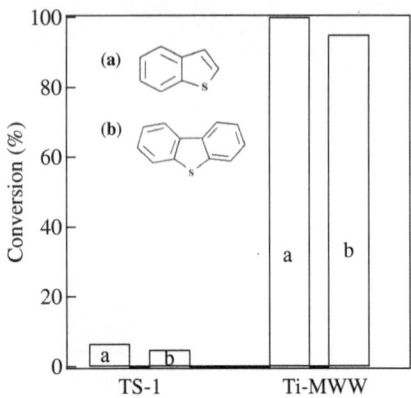

polarity by selective oxidation, and extraction with suitable solvents are conducted to realize the desulfurization [84].

Titanosilicate/H_2O_2 catalytic systems such as Ti-MWW/H_2O_2 and TS-1/H_2O_2 have been used to the oxidative desulfurization process for model light oil [39, 85]. As shown in Fig. 4.19, when the catalytic reactions were carried out in acetonitrile at 343 K, Ti-MWW showed the conversion of benzothiophene and dibenzothiophene both >95 %, while TS-1 showed trace conversion for both substrates. Benothiophene and dibenzothiophene are oxidized to their corresponding sulfoxides and sulfones, which can be removed readily by extraction with acetonitrile.

The advantages of Ti-MWW were found in the oxidative desulfurization of bulky aromatic sulfur compounds like 4,6-dimethyldibenzothiophene (4,6-DMDBT) [39]. Table 4.19 compares the oxidation activity for various sulfur compounds over TS-1, Ti-Beta, Ti-MCM-41, and Ti-MWW. With increasing molecular dimension of the substrates, the conversion of sulfur compound became low. Obviously, Ti-MWW was the most effective catalyst for oxidative desulfurization of 4,6-DMDBT. This is attributed to the contribution of the 12-MR side cups in Ti-MWW, which are easily accessible to the bulky molecules.

4.5.4 Epoxidation and Cyclization of Olefinic Alcohols

As shown, Ti-MWW efficiently catalyzes the epoxidation of AAL and diallyl ether with H_2O_2 as oxidant. The main products were corresponding epoxides [28, 29]. In the case of the epoxidation of the unsaturated alcohols with a longer chain (ω-hydroxyalkenes with more than C_5), however, the cyclization of the epoxides takes place possibly, leading to corresponding hydroxytetrahydrofuran or hydroxytetrahydropyran (Scheme 4.15). These 5-MR or 6-MR products are produced via two steps. Firstly, the double bond of olefinic alcohols is activated to the corresponding unstable epoxy alcohols by electrophilic attack of the titanium

Table 4.19 Oxidation of aromatic organic sulfur compounds over various titanosilicates[a]. Reprinted from Ref. [39], Copyright 2010, with permission from John Wiley and Sons

Entry	Catalyst	Sulfur compound[b]	Conversion (%)
1	TS-1	TH	41
		BT	19
		DBT	4
		4,6-DMDBT	1
2	Ti-Beta	TH	4
		BT	100
		DBT	84
		4,6-DMDBT	35
3	Ti-MCM-41	TH	2
		BT	87
		DBT	94
		4,6-DMDBT	35
4	Ti-MWW	TH	36
		BT	87
		DBT	84
		4,6-DMDBT	78

[a] Reaction conditions: TS-1 or Ti-MWW, 40 mg; Ti-Beta, 48 mg; Ti-MCM-41, 47 mg; model light oil (100 mg mL^{-1} S), 10 mL; solvent (water for TS-1, MeCN for the others), 10 mL, 323 K, 3 h
[b] TH, thiophene; BT, benzothiophene; DBT, dibenzothiophene; 2,4-DMDBT, 2,4-dimethyl dibenzothiophene

hydroperoxo oxygen over titanoilicate. Secondly, the epoxides undergo cyclization by the intramolecular nucleophilic attack of terminal hydroxyl moiety present in the molecule to the activated carbon atom of the epoxy ring. This step occurs on the weak acid sites of titanosilicates. Taking advantages of this kind of ring closure of substituted 4-penten-1-oxy and 5-hexen-1-oxy radicals, it is possible to synthesize tetrahydrofuran and tetrahydropyran ring compounds in one-pot. The widespread occurrence of these substituted compounds in many classes of natural products have made them valuable building blocks for the synthesis of various biologically active organic target molecules. We found Ti-MWW could serve as a promising catalyst for synthesizing these functional compounds.

As shown in Scheme 4.15, the cyclization of 4-penten-1-ol occured regiose-lectively to produce the 5-exo product tetrahydrofurfuryl alcohol (5-MR). On the other hand, the corresponding tetrahydropyran regioisomer of 5-MR supposed to form via endo attack of the hydroxy nucleophile to the activated terminal carbon atom of the epoxide intermediate takes the form of tetrahydro-4H-pyran-3-ol (6-MR). This is known as Baldwin rule. According to this rule, exo nucleophilic attack occurs mainly. However, as given in Table 4.20, this 6-MR product was not formed at all over TS-1 because its 10-MR medium pores limit the formation of bulky products. Nevertheless, tetrahydro-4H-pyran-3-ol (6-MR) was synthesized with a yield about 2–3 % over Ti-Beta and Ti-MWW both with a larger pore size.

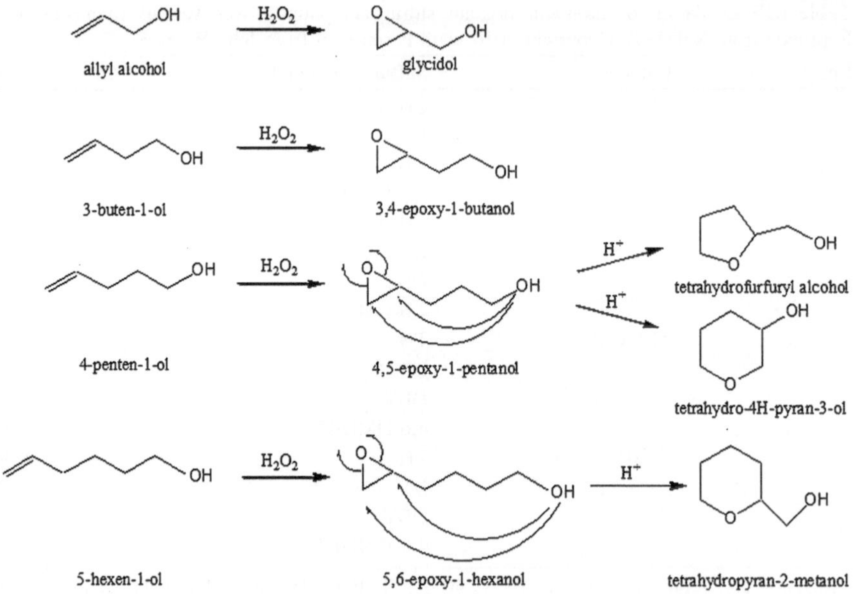

Scheme 4.15 Product distribution in the epoxidation of various unsaturated alcohols

In the case of 5-hexen-1-ol, two products, either tetrahydropyran or 7-MR redioisomer tetrahydrohomopyran derivative, are supposed to be produced, but only tetrahydropyran-2-netanol was formed. Here the exo attack prevails to give 6-MR tetrahydropyran rather than the 7-MR regioisomer, which is formed via endo attack. Although the conversion and the specific activity (turnover frequency, TOF) decreased with increasing the carbon chain of the olefinic alcohol for three type of titanosilicates, Ti-MWW was obviously more active than TS-1 and Ti-Beta. This should be attributed to a high accessibility of the Ti active sites to the linear alcohol molecules inside the 10-MR channels of the MWW topology. Meanwhile, its 12-MR supercages and side cups could accommodate cyclic compounds, which allow the cyclization of the epoxides to occur possibly. The unique structure of Ti-MWW is thus helpful for the epoxidation and the subsequent cyclization.

4.6 Selective Epoxidation of Propylene to PO with H_2 and O_2 Over Bifunctional Au/Ti-MWW Catalyst

Ti-MWW is capable of catalyzing the epoxidation reaction of propylene to produce PO with H_2O_2 as oxidant in the liquid phase [33]. In recent years, many efforts have been devoted to the direct epoxidation of propylene with H_2 and O_2

Table 4.20 The results of the epoxidation of various unsaturated alcohols[a]

Catalyst	Substrate	Time/h	Conversion/%	Selectivity/%			TOF/h^{-1}
				Epoxide	5-MR	6-MR	
TS-1 (Si/Ti = 40)	Allyl alcohol[b]	1	11.8	100	–	–	117
	3-buten-1-ol	2.5	21.8	100	–	–	43
	4-penten-1-ol	2.5	21.1	3.1	96.9	0	42
	5-hexen-1-ol	2.5	16.8	36.0	–	63.9	33
Ti-MWW (Si/Ti = 42)	Allyl alcohol[b]	1	63.5	100	–	–	670
	3-buten-1-ol	2.5	59.0	100	–	–	124
	4-penten-1-ol	2.5	64.0	15.0	83.0	2.0	135
	5-hexen-1-ol	2.5	24.0	43.1	–	56.9	23
Ti-Beta (Si/Ti = 40)	Allyl alcohol[b]	1	7.8	100	–	–	77
	3-buten-1-ol	2.5	13.2	100	–	–	26
	4-penten-1-ol	2.5	25.3	0	97.0	3.0	50
	5-hexen-1-ol	2.5	8.8	17.5	–	82.5	18

[a] Reaction conditions: cat., 50 mg; reaction temp., 333 K; solvent, MeCN, 10 mL; substrate, 5 mmol; H_2O_2 (31 wt %), 5 mmol
[b] cat., 25 mg

over Au/titanium-containing support. Haruta and co-workers first reported in 1998 that Au/anatase catalysts prepared by deposition–precipitation (DP) method could catalyze the propylene epoxidation with H_2/O_2 at a PO selectivity >95 % and a low propylene conversion (~1 %) [86]. So far, considerable research has been carried out in this area, including the investigation of Au supported on titanosilicates [87–89]. Delgass and co-workers found that a stable catalyst of 0.05 wt % Au/TS-1 (Si/Ti = 36) gave a propylene conversion of 8.8 % and a PO selectivity of 81 % at 473 K. It is believed that H_2 and O_2 react on Au nanoparticles to form H_2O_2, which then migrates to Ti sites to epoxidize propylene adsorbed there [89]. It is shown that the hydroperoxide species on the Ti site are the active species for epoxidation.

Thus, Au nanoparticles have been supported on Ti-MWW by DP method, leading to a bifunctional catalyst Au/Ti-MWW (Scheme 4.16). Its catalytic performance has been investigated in the direct epoxidation of propylene with H_2 and O_2 [40]. The catalytic results of Au/Ti-MWW have been compared with Au/TS-1. The TEM images showed the size of Au nanoparticles observed on the fresh Au/Ti-MWW and Au/TS-1 was both smaller than 5 nm. Considering the fact that the pore entrance of TS-1 and Ti-MWW was only about 5.5 Å, the Au nanoparticles observed were believed on the outer surface of zeolite crystallites. The Au loading increased slightly with increasing Ti content both in Ti-MWW and TS-1, which was consistent with the results reported by other groups that higher Ti content led to higher Au loading on TS-1 zeolite [90]. It seems that there is a certain affinity interaction between Ti species and Au precursor. The presence of Ti species is

Scheme 4.16 Bifunctional
Au/Ti-MWW catalyst for
direct epoxidation of
propylene with H_2 and O_2.
Reprinted from Ref. [40],
Copyright 2012, with
permission from International
Union of Pure and Applied
Chemistry

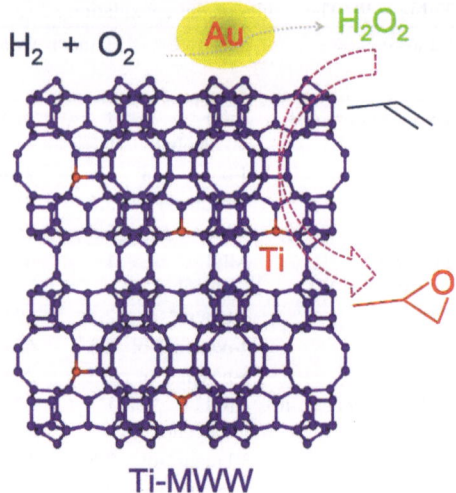

helpful for capturing the Au precursor in the DP process. Moreover, compared with that at the same Si/Ti ratio, the Au content supported on Ti-MWW was higher than that on TS-1, which is probably due to more defect sites contained in Ti-MWW framework.

The Au/TS-1 and Au/Ti-MWW with various Ti contents and Au loading were applied to the gas-phase epoxidation of propylene with H_2 and O_2 [40]. The reaction gave the products including CO, CO_2, H_2O, and C_3 compounds including PO, propane, propionaldehyde, acetone, and acrolein. The catalytic activities of catalysts were clearly related to the Au loading and Si/Ti ratio as given in Table 4.21. The initial activity was in the order of Au/Ti-MWW(40) > Au/Ti-MWW(70) > Au/Ti-MWW(135) > Au/Ti-MWW(165) > Au/Ti-MWW(225) > Au/Ti-MWW(350). The lower the Si/Ti ratio is, the lower the Au content and initial activity are. The similar results were also observed in the case of Au/TS-1. It is believed that H_2 and O_2 react on Au nanoporticles to form H_2O_2, which then migrates to Ti sites to epoxidize propylene adsorbed. Thus, it is comprehensible that more Au nanoparticles would promote in situ generation of H_2O_2 and then favor the epoxidation of propylene on Ti sites. In addition to the propylene conversion, the PO selectivity is also important. The selectivity of PO increased with Si/Ti ratio of Ti-MWW supports. However, Au/Ti-MWW not only showed lower propylene conversion, but also lower PO selectivity than Au/TS-1. With comparable Si/Ti ratios, Au/Ti-MWW(135) and Au/TS-1(140) showed the initial PO selectivities of 39.6 and 89.0 %, respectively. The deep oxidation products (mainly CO_2 together with CO) were produced at a higher level on Au/Ti-MWW than Au/TS-1.

There are abundant defects in the framework of Ti-MWW zeolite due to the Ti-MWW preparation method as we discussed before [14]. The Si–OH groups in the

Table 4.21 The results of propylene epoxidation over Au/Ti-MWW and Au/TS-1 with various Si/Ti ratios[a]. Reprinted from Ref. [40], Copyright 2012, with permission from International Union of Pure and Applied Chemistry

Catalyst	Si/Ti ratio	Au (wt %)	C_3H_6 conversion (%)[a]	Product selectivity[b] (%)						PO formation rate (g kg^{-1} h^{-1})
				PO	Pn	PA	Ac	An	CO + CO$_2$	
Au/Ti-MWW(40)	40	0.28	5.1(1.5)	13.4(13.3)	13.0(15.9)	16.6(14.7)	13.1(8.7)	26.0(4.0)	16.0(39.5)	7.1(2.1)
Au/Ti-MWW(70)	70	0.24	3.1(2.0)	12.0(22.9)	10.8(13.0)	16.4(13.4)	7.1(6.2)	2.2(2.9)	46.9(39.5)	3.9(4.7)
Au/Ti-MWW(135)	135	0.19	2.5(2.1)	39.6(35.8)	9.3(11.6)	5.7(6.6)	5.3(6.7)	3.1(3.0)	37.0(36.2)	10.1(7.4)
Au/Ti-MWW(165)	165	0.16	1.5(1.4)	44.2(46.9)	12.0(12.7)	9.3(8.0)	5.7(4.9)	1.6(0)	27.2(27.6)	6.8(6.9)
Au/Ti-MWW(225)	225	0.13	1.5(1.4)	57.1(63.2)	12.1(11.2)	4.4(4.4)	3.9(3.3)	0(0)	22.6(17.8)	9.1(9.4)
Au/Ti-MWW(350)	350	0.1	1.4(1.3)	58.1(59.4)	13.7.6(9.0)	12.3(12.1)	7.8(9.7)	0(0)	8.2(9.8)	8.1(7.8)
Au/TS-1(35)	35	0.17	7.1(6.6)	33.7(37.2)	5.6(6.2)	8.4(8.8)	4.0(5.5)	1.3(1.6)	42.9(35.2)	24.8(25.3)
Au/TS-1(70)	70	0.12	4.6(2.1)	73.2(88.5)	5.2(0)	6.9(5.5)	2.9(0)	0(0)	11.9(6.0)	34.7(19.0)
Au/TS-1(140)	140	0.08	3.6(2.2)	89.0(88.0)	2.7(5.9)	2.3(2.0)	0(0)	0(0)	6.0(4.2)	32.8(20.0)
Au/TS-1(240)	240	0.05	3.1(2.4)	92.9(91.5)	0(0)	2.9(3.3)	0(0)	0(0)	4.2(5.2)	29.6(22.6)

[a] Reaction conditions: C_3H_6:H_2:O_2:N_2 = 1:1:1:7; space velocity = 4000 mL g^{-1} h^{-1}; temperature = 423 K. The data out of and within parentheses were observed at time on stream of 0.5 and 5 h, respectively

[b] PO propene oxide. Pn propane. PA propionaldehyde. Ac acetone. An acrolein

framework of Ti-MWW would adsorb PO efficiently. Thus, the desorption of PO from Au/Ti-MWW was not as easy as from Au/TS-1, leading to a lower PO selectivity on Au/Ti-MWW. To enhance the PO selectivity of Au/Ti-MWW, it is necessary to improve its hydrophobicity. We once developed a post structural rearrangement technique assisted by PI treatment that proved to be effective for reducing the amount of Si–OH groups in Ti-MWW [74]. Thus, Ti-MWW was first rearranged with PI, giving rise to Re–Ti-MWW. A part of defect sites (hydroxyl nests) were removed during this structural rearrangement. Both IR and ^{29}Si MAS NMR investigations verified the decrease in hydroxyl groups after PI treatment [40]. Re–Ti-MWW was then supported with Au nanoparticles to investigate the catalytic performance in direct epoxidation of propylene to PO with H_2 and O_2. The PO selectivity was significantly improved after PI treatment. Figure 4.20 shows the dependence of propylene conversion and PO selectivity on the Ti content of Au/TS-1, Au/Ti-MWW, and Au/Re–Ti-MWW. The PI rearrangement lowered slightly the catalytic activity for propylene conversion, but greatly enhanced the PO selectivity. Thus, Au/Re–Ti-MWW showed almost the same PO selectivity dependence on Ti content as TS-1. This is clearly related to the reduction of Si–OH groups in defect sites by post-treatment of Ti-MWW with PI and further calcination. The reduction of Si–OH groups would speed up the desorption of PO from zeolite surface and channels. This would minimize the consecutive side reactions and improve the selectivity to the objective product of PO. The highest initial PO formation rate among the series of Au/Re–Ti-MWW catalysts was obtained on Au/Re–Ti-MWW(135). The value of 22 g_{po} kg^{-1} h^{-1} was about twice as high as that achieved on Au/Ti-MWW(135).

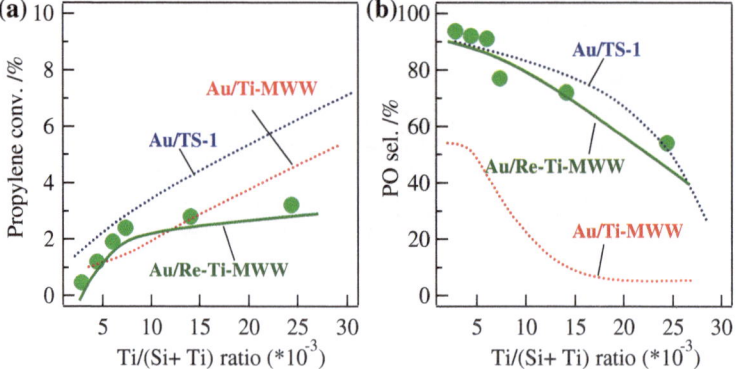

Fig. 4.20 The propylene conversion (**a**) and PO selectivity (**b**) as a function of Ti content in Au/Re–Ti-MWW. For comparison, the results of Au/Ti-MWW and Au/TS-1 are shown in dash lines. Reprinted from Ref. [40], Copyright 2012, with permission from International Union of Pure and Applied Chemistry

References

1. Taramasso T, Perego G, Notari B (1983) US Patent 4,410,50
2. Vayssilov GN (1997) Structural and physicochemical features of titanium silicates. Catal Rev Sci Eng 39:209–244
3. Gao X, Wachs IE (1999) Titania-silica as catalysts: molecular structural characteristics and physic-chemical properties. Catal Today 51:233–254
4. Ratnasamy P, Srinivas D, Knözinger H (2004) Active sites and reactive intermediates in titanium silicate molecular sieves. Adv Catal 48:1–169
5. Petrini G, Leofanti G, Mantegazza MA, Pignataro F (1996) Green chemistry: designing chemistry for the environment. ACS symposium series. American Chemical Society, Washington DC, p 33
6. Perego C, Carati A, Ingallina P, Mantegazza MA, Bellussi G (2001) Production of titanium containing molecular sieves and their application in catalysis. Appl Catal A: General 221:63–72
7. Cavani F, Teles JH (2009) Sustainability in catalytic oxidation: an alternative approach or a structural evolution? ChemSusChem 2:508–534
8. Reddy JS, Kumar R, Ratnasamy P (1990) Titanium silicate-2: synthesis, characterization and catalytic properties. Appl Catal 58:L1–L4
9. Corma A, Camblor MA, Esteve P, Martinez A, Perezpariente J (1994) Activity of Ti-Beta catalyst for the selective oxidation of alkenes and alkanes. J Catal 145:151–158
10. Camblor MA, Costantini M, Corma A, Gilbert L, Esteve P, Martinez A, Valencia S (1996) Synthesis and catalytic activity of aluminium-free zeolite Ti-beta oxidation catalysts. Chem Commun 11:1339–1340
11. Tuel A (1995) Synthesis, characterization, and catalytic properties of the new TiZSM-12 zeolite. Zeolites 15:236–242
12. Wu P, Komatsu T, Yashima T (1997) Ammoximation of ketones over titanium mordenite. J Catal 168:400–411
13. Diaz-Cabanas M, Villaescusa LA, Camblor MA (2000) Synthesis and catalytic activity of Ti-ITQ-7: a new oxidation catalyst with a three-dimensional system of large pore channels. Chem Commun 9:761–762
14. Wu P, Tatsumi T, Komatsu T, Yashima T (2001) A novel titanosilicate with MWW structure. I. Hydrothermal synthesis, elimination of extraframework titanium, and characterizations. J Phys Chem B 105:2897–2905
15. Tanev PT, Chibwe M, Pinnavaia TJ (1994) Titanium-containing mesoporous molecular sieves for catalytic oxidation of aromatic compounds. Nature 368:321–323
16. Newalkar BL, Olanrewaju J, Komarneni S (2001) Direct synthesis of titanium-substituted mesoporous SBA-15 molecular sieve under microwave-hydrothermal conditions. Chem Mater 13:552–557
17. Leonowicz ME, Lawton JA, Lawton SL, Rubin MK (1994) MCM-22: a molecular sieve with two independent multidimensional channel systems. Science 164:1910–1913
18. Lawton SL, Leonowicz ME, Partridge PD, Chu P, Rubin MK (1998) Twelve-ring pockets on the external surface of MCM-22 crystals. Microporous Mesoporous Mater 23:109–117
19. Wu P, Nuntasri D, Ruan JF, Liu YM, He MY, Fan WB, Terasaki O, Tatsumi T (2004) Delamination of Ti-MWW and high efficiency in epoxidation of alkenes with various molecular sizes. J Phys Chem B 108:19126–19131
20. Wang LL, Liu YM, Xie W, Wu HH, Jiang YW, He MY, Wu P (2007) Postsynthesis and catalytic properties of metallosilicates structurally analogous to MCM-56. Stud Surf Sci Catal 170:635–640
21. Kim S, Ban H, Ahn W (2007) Ti-MCM-36: a new mesoporous epoxidation catalyst. Catal Lett 113:160–164

22. Fan WB, Wu P, Namba S, Tatsumi T (2004) A titanosilicate that is structurally analogous to an MWW-type lamellar precursor. Angew Chem Int Ed 43:236–244

23. Ruan JF, Wu P, Slater B, Terasaki O (2005) Structure elucidation of the highly active titanosilicate catalyst Ti-YNU-1. Angew Chem Int Ed 44:6719–6723

24. Wu P, Ruan JF, Wang LL, Wu LL, Wang Y, Liu YM, Fan WB, He MY, Terasaki O, Tatsumi T (2008) Methodology for synthesizing crystalline metallosilicates with expended pore windows through molecular alkoxysilylation of zeolitic lamellar precursors. J Am Chem Soc 130:8178–8187

25. Wang LL, Wang Y, Liu YM, Wu HH, Li XH, He MY, Wu P (2009) Alkoxysilylation of Ti-MWW lamellar precursors into interlayer pore-expanded titanosilicates. J Mater Chem 19:8594–8602

26. Wu P, Tatsumi T, Komatsu T, Yashima T (2001) A novel titanosilicate with MWW structure: II. Catalytic properties in selective oxidation of alkenes. J Catal 202:245–255

27. Wu P, Tatsumi T (2002) Unique *trans*-selectivity of Ti-MWW in epoxidation of *cis/trans*-alkenes with hydrogen peroxide. J Phys Chem B 106:748–753

28. Wu P, Tatsumi T (2003) A novel titanosilicate with MWW structure III. Highly efficient and selective production of glycidol through epoxidation of allyl alcohol with H_2O_2. J Catal 214:317–326

29. Wu P, Liu YM, He MY, Tatsumi T (2004) A novel titanosilicate with MWW structure: catalytic properties in selective epoxidation of diallyl ether with hydrogen peroxide. J Catal 228:183–191

30. Wu P, Nuntasri D, Liu YM, Wu HH, Jiang YW, Fan WB, He MY, Tatsumi T (2006) Selective liquid-phase oxidation of cyclopentene over MWW type titanosilicate. Catal Today 117:199–205

31. Wang LL, Liu YM, Xie W, Zhang HJ, Wu HH, Jiang YW, He MY, Wu P (2007) Highly efficient and selective production of epichlorohydrin through epoxidation of allyl chloride with hydrogen peroxide over Ti-MWW catalysts. J Catal 246:205–214

32. Wu HH, Liu YM, Wang LL, Zhang HJ, He MY, Wu P (2007) Epoxidation of 2,5-dihydrofuran to 3,4-epoxytetrahydrofuran over Ti-MWW catalysts. Appl Catal A: General 320:173–180

33. Song F, Liu YM, Wang LL, Zhang HJ, He MY, Wu P (2007) Highly efficient epoxidation of propylene over a novel Ti-MWW catalyst. Studies In Surface Science Catalysis 170:1236–1243

34. Song F, Liu YM, Wu HH, He MY, Wu P, Tatsumi T (2006) A novel titanosilicate with MWW stucuture: highly effective liquid-phase ammoximation of cyclohexanone. J Catal 237:359–367

35. Song F, Liu YM, Wang LL, Zhang HJ, He MY, Wu P (2007) Highly selective synthesis of methyl ethyl ketone oxime through ammoximation over Ti-MWW. Appl Catal A: General 327:22–31

36. Zhao S, Xie W, Yang J. Liu YM, Zhang YT, Xu BL, Jiang JG, He MY, Wu P (2011) An investigation into cyclohexanone ammoximation over Ti-MWW in a continuous slurry reactor. Appl Catal A: General 394:1–8

37. Xu L, Peng HG, Zhang K, Wu HH, Chen L, Liu YM, Wu P (2013) Core-shell-structured titanosilicate as a robust catalyst for cyclohexanone ammoximation. ACS Catal 3:103–110

38. Xie W, Zheng YT, Zhao S, Yang J, Liu YM, Wu P (2010) Selective oxidation of pyridine to pyridine-N-oxide with hydrogen peroxide over Ti-MWW catalyst. Catal Today 157:114–118

39. Gao GH, Cheng SF, An Y, Si XJ, Fu XL, Liu YM, Zhang HJ, Wu P, He MY (2010) Oxidative desulfurization of aromatic sulfur compounds over titanosilicates. ChemCatChem 2:459–466

40. Ren YJ, Xu L, Zhang LY, Wang JG, Liu YM, He MY, Wu P (2012) Selective epoxidation of propylene to propylene oxide with H_2 and O_2 over Au/Ti-MWW catalysts. Pure Appl Chem 84:561–578

41. Bellussi G, Rigguto MS (1994) Metal ions associated to the molecular sieve framework: possible catalytic oxidation sites. Stud Surf Sci Catal 85:177–213
42. Notari B (1996) Microporous crystalline titanium silicates. Adv Catal 41:253–334
43. Koyano KA, Tatsumi T (1996) Synthesis of titanium-containing mesoporous molecular sieves with a cubic structure. Chem Commun 2:145–146
44. Cavani F, Teles JH (2009) Sustainability in catalytic oxidation: an alternative approach or a structural evolution? ChemSusChem 2:508–534
45. Nijhuis TA, Makkee M, Moulijn JA, Weckhuysen BM (2006) The production of propene oxide: catalytic processes and recent development. Ind Eng Chem Res 45:3447–3459
46. Watcher, Kirk-Othmer (1998) Encyclopedia of chemical technology, 4th edn. Wiley, New York, pp 137–141
47. Teles JH, Rehfinger A, Bassler P, Wenzal A, Rieber N, Rudolf P (2004) US Patent 6,756,503
48. Clerici MG, Ingallina P (1993) Epoxidation of lower olefins with hydrogen peroxide and titanium silicate. J Catal 140:71–83
49. Wu P, Tatsumi T (2001) Extremely high trans selectivity of Ti-MWW in epoxidation of alkenes with hydrogen peroxide. Chem Commun 10:897–898
50. Wittcoff HA, Reuben BG (1996) Industrial organic chemicals. Wiley, New York, p 189
51. Hutchings GJ, Lee DF (1994) Oxidation of thioethers and sulfoxides with hydrogen peroxide using TS-1 as catalyst. J Chem Soc Chem Commun 1095–1096
52. Hutchings GJ, Lee DF, Minihan AR (1995) Epoxidation of allyl alcohol to glycidol using titanium silicate TS-1: effect of the method of preparation. Catal Lett 33:369–385
53. Hutchings GJ, Lee DF, Minihan AR (1996) Epoxidation of allyl alcohol to glycidol using titanium silicate TS-1: effect of the reaction conditions and catalyst acidity. Catal Lett 39:83–90
54. Wittcoff HA, Reuben BG (1996) Industrial organic chemicals. Wiley, New York, p 187
55. Bellussi G, Carati A, Clerici MG, Maddinelli G, Millini R (1992) Reactions of titanium silicate with protic molecules and hydrogen peroxide. J Catal 133:220–230
56. Imura S, Ohtsuru M (1997) JP Patent 5,214,7699
57. Hodgson DM, Stent MAH, Wilson FX (2001) Substituted alkenediols by alkylative double ring opening of dihydrofuran and dihydropyran epoxides. Org Lett 3:3401–3403
58. Hodgson DM, Stent MAH, Stefane B, Wilson FX (2003) Enantioselective alkylative double ring-opening of epoxides derived from cyclic allylic ethers: synthesis of enantiorn riched unsaturated diols. Org Biomol Chem 7:1139–1150
59. Lai GF (2004) A convenient preparation of tetrahydrofuran-based diamines. Synth Commun 34:1981–1987
60. Martinez LE, Leighton JL, Carsten DH, Jacobson EN (1995) Highly enantioselective ring opening of epoxides catalyzed by (salen)Cr(III) complexes. J Am Chem Soc 117:5897–5898
61. Barili PL, Berti G, Mastrorili E (1993) Regio- and stereochemistry of the acid catalyzed and of a highly enantiselective enzymatic hydrolysis of some epoxytetrahydrofurans. Tetrahedron 49:6263–6276
62. Kim WJ, Kim TJ, Ahn WS, Lee YJ, Yoon KB (2003) Synthesis, characterization and catalytic properties of TS-1 monoliths. Catal Lett 91:123–127
63. Wu HH, Wang LL, Zhang HJ, Liu YM, Wu P, He MY (2006) Highly efficient and clean synthesis of 3,4-epoxytetrahydrofuran over a novel titanosilicate catalyst Ti-MWW. Green Chem 8:78–81
64. Gao HX, Lu WK, Chen QL (2002) Reaction kinetics of epoxidation of allyl chloride with hydrogen peroxide catalyzed by titanium silicate-1. Chin J Catal 23:3–9
65. Gao HX, Lu GX, Suo JS, Li SB (1996) Epoxidation of allyl chloride with hydrogen peroxide catalyzed by titanium silicate-1. Appl Catal A: General 138:27–38
66. Wu P, Tatsumi T (2002) Preparation of B-free Ti-MWW through reversible structural conversion. Chem Commun 10:1026–1027
67. Bellussi G, Rigguto MS (2001) Chapter 19 metal ions associated to molecular sieve frameworks as catalytic sites for selective oxidation reactions. Stud Surf Sci Catal 137:911–955

68. Ichihashi H, Sato H (2001) The development of new heterogeneous catalytic processes for the production of ε-caprolactam. Appl Catal A:General 221:359–366
69. Roffia P, Padovan M, Leofanti G, Mantegazza MA, Alberti GD, Tauszik GR (1998) US Patent 4,794,198
70. Zhao S, Xie W, Liu YM, Wu P (2011) Methyl ethyl ketone ammoximation over Ti-MWW in a continuous slurry reactor. Chin J Catal 32(1):179–183
71. Thangaraj A, Sivasanker S, Ratnasamy P (1991) Catalytic properties of crystalline titanium silicates III. Ammoximation of cyclohexanone. J Catal 131:394–400
72. Roffia P, Leofanti G, Cesana A, Mantegazza M, Padovan M, Petrini G, Tonti S, Gervasutti P (1990) Cyclohexanone ammoximation: a breakthrough in the 6-caprolactam production process. Stud Surf Sci Catal 55:43–52
73. Zecchina A, Spoto G, Bordiga S, Geobaldo F, Petrini G, Leofanti G, Padovan M, Mantegazza M, Roffia P (1993) Ammoximation of cyclohexanone on titanium silicate: investigation of the reaction mechanism. Stud Surf Sci Catal 75:719–729
74. Wang LL, Liu YM, Xie W, Wu HH, Li XH, He MY, Wu P (2008) Improving the hydrophobicity and oxidation activity of Ti-MWW by reversible structural rearrangement. J Phys Chem C 112:6132–6138
75. Fang XQ, Wang Q, Zheng AM, Liu YM, Wang YN, Deng XJ, Wu HH, Deng F, He MY, Wu P (2012) Fluorine-planted titanosilicate with enhanced catalytic activity in alkene epoxidation with hydrogen peroxide. Catal Sci Tech 2:2433–2435
76. Fang XQ, Wang Q, Zheng AM, Liu YM, Lin LF, Wu HH, Deng F, He MY, Wu P (2013) Post-synthesis, characterization and catalytic properties of fluorine-planted MWW-type titanosilicate. Phys Chem Chem Phys 15:4930–4938
77. Malkov AV, Bell M, Castelluzzo F, Kocovsky P (2005) Methox: a new pyridine-N-oxide organocatalyst for the asymmetric allylation of aldehydes with allyltrichlorosilanes. Org Lett 7:3219–3222
78. Bockelhe V, Linn WJ (1954) Rearrangements of N-oxide. a novel synthesis of pyridylcarbinols and aldehydes. J Am Chem Soc 76:1286–1291
79. Adama CD, Scanian PA, Secrist ND (1994) Oxidation and biodegradability enhancement of 1,4-dioxane using hydrogen peroxide and ozone. Environ Sci Technol 28:1812–1818
80. Son HS, Choi SB, Khan E, Zoh KD (2006) Removal of 1,4-dioxane from water using sonication: effect of adding oxidations on the degradation kinetics. Water Res 40:692–698
81. Suh JH, Mohseni M (2004) A study on the relationship between biodegradability enhancement and oxidation of 1,4-dioxane using ozone and hydrogen peroxide. Water Res 38:2596–2604
82. Fan WB, Kubota Y, Tatsumi T (2008) Oxidation of 1,4-dioxane over Ti-MWW in the presence of H_2O_2. ChemSusChem 1:175–178
83. Forni L, Bahnemann D, Hart EJ (1982) Mechanism of the hydroxide ion-initiated decomposition of ozone in aqueous solution. J Phys Chem 86:255–259
84. Li C, Jiang ZX, Gao JB, Yang YX, Wang SJ, Tian FP, Sun FX, Sun XP, Ying PL, Han CR (2004) Ultra-deep desulfurization of diesel: oxidation with a recoverable catalyst assembled in emulsion. Chem-A Euro J 10:2277–2280
85. Cheng SF, Liu YM, Gao JB, Wang LL, Liu XL, Gao GH, Wu P, He MY (2006) Catalytic oxidation of benzothiophene and dibenzothiophene in model light oil over Ti-MWW. Chin J Catal 27:547–549
86. Hayashi T, Tanaka K, Haruta M (1998) Selective vapor-phase epoxidation of propylene over Au/TiO2 catalysts in the presence of oxygen and hydrogen. J Catal 178:566–575
87. Qi CX, Akita T, Okumura M, Haruta M (2001) Epoxidation of propylene over gold catalysts supported on non-porous silica. Appl Catal A: General 218:81–89
88. Yap N, Andres RP, Delgass WN (2004) Reactivity and stability of Au in and on TS-1 for epoxidation of propylene with H_2 and O_2. J Catal 226:156–170

89. Cumaranatunge L, Delgass WN (2005) Enhancement of Au capture efficiency and activity of Au/TS-1 catalysts for propylene epoxidation. J Catal 232:38–42
90. Taylor B, Lauterbach J, Delgass WN (2005) Gas-phase epoxidation of propylene over small gold ensembles on TS-1. Appl Catal A: General 291:188–198

Chapter 5
Conclusions and Prospects

Selective oxidation is one of the most important reactions, which serves as a bridge between basic raw materials from fossil resource and the oxyfunctional products such as polymer, fine chemicals, pharmaceuticals, and agricultural chemicals. The applications range from laboratory scale organic synthesis to commercial production of bulk chemicals in petrochemical industry. Great concerns have risen to those conventional noncatalytic oxidation processes suffering serious problems of environmental pollution and byproduct disposal. Providing a useful substitution for developing greener oxidation processes, the titanosilicate/H_2O_2 system has already achieved great successes in the production of cycohexanone oxime and propylene oxide. Greener processes and clean production of specific oxygenated chemicals are waiting for tailor-made titanosilicate catalysts matching up well with reactants and products. As one of the representative titanosilicates developed in the past few decades, layered Ti-MWW titanosilicate has achieved great progress in synthesis, structural modification, and catalytic applications, which is reviewed comprehensively in this book. The ideas would be of some importance to develop other novel oxidation catalysts. In addition to developing new materials, design and synthesis of titanosilicat-based bifunctional or multifunctional catalysts are expected to bring about more attractive oxidation processes, not using expensive hydrogen peroxide but employing hydrogen and oxygen, or more preferable oxidant of oxygen. Those catalysts applicable to tandem reactions are also desirable for one-pot synthesis of the chemicals that usually require multiple steps.